SA's Green Treatment: Aeration Saves Energy

Vani

CONTENTS

1. INTRODUCTION

According to The Carbon Report, South Africa currently ranks among the dirtiest energy producers in the world due to its inherent reliance on coal. South Africa's carbon emissions of 8.9 tonnes per capita per year on average is among the highest in the developing world (The Carbon Report- Why is this relevant to South African Companies, 2015). With the South African government's push to reduce carbon emissions by 34% by 2020 and 42% by 2025 and South Africa's sole energy producer, Eskom, currently on R220.8 billion of guarantees from National Treasury, (Eskom a risk to South Africa as finances are getting worse, 2018) and a debt of nearly R 0.5 trillion, electricity cost in South Africa is set to increase rapidly over the next few years.

Although energy consumption per capita for wastewater treatment is low (<5 W/person), electricity consumption at wastewater treatment plants (WWTPs) is high due to the large number of people connected to the WWTP, running onto some 1MW for a 25Mℓ/d WWTP. So, as WWTPs in South Africa are large energy consumers, lowering of energy consumption at WWTPs is becoming more crucial. It is therefore important that all measures be taken to optimise and conserve energy at WWTPs. As aeration accounts by far for the most energy consumed on a WWTP, between 45 and 60% (Roman & Mureşan, 2014), choosing the correct aeration technology and implementing the correct aeration process control regimes are the most important choices that can be made when designing a WWTP.

Musvoto and Ikumi (2016) categorize measures to conserve energy (what they call Energy Conservation Measures (ECMs)) on WWTPs into different groups. These ECMs are listed in descending order of ease of implementation and capital investment requirements (i.e. the easiest and least cost first) as detailed below:

1) Simple measures that only require changes to process operation and control to optimal levels, with little to no additional capital investment apart from operator training.

2) Low to medium capital measures that involve upgrading aeration and control strategies requiring investment in new monitoring equipment and control systems.

3) Complex measures that involve (i) redesigning and replacing less efficient aeration systems with more efficient technologies (ii) introduction of influent flow balancing and/or implementing anaerobic digestion for energy recovery. (Musvoto & Ikumi, 2016)

From these three categories, categories (1) and (2) are simple strategies and require low to medium capital investment to implement. For this reason, these strategies are currently being implemented on various WWTPs in South Africa. However, looking at category (3), where larger capital investment is required, the implementation is far less common. With this in mind, this study will focus on the "redesigning and replacing of less efficient aeration systems (slow speed surface aeration) with more efficient technologies (fine bubble diffused aeration)". As with any large capital investment project, the viability of the investment must be investigated. Looking at information readily available in literature, various sources could be found that compares the difference in efficiencies between the two different technologies (SSSA vs FBDA) but none that compares overall costs (capital, maintenance and operational) for the two different technologies. It would seem there is a lack of information with regards overall financial costing of these different aeration technologies, making viability calculations very uncertain. Investors and designers are therefore unsure as to which of the aeration technologies would be the most viable option.

For this reason, this study will focus on costing models (capital, operation and maintenance) for these two different aeration technologies i.e. Slow Speed Surface Aeration (SSSA) and Fine Bubble Diffused Aeration (FBDA). Two different process configurations i.e. the Modified Ludzack-Ettinger (MLE) and University of Cape Town (UCT), will be investigated together with different inflow rates (in Mℓ/d) and different inflow wastewater characteristics to test the viability of these systems. The goal of the study is to determine the inflow rate and wastewater inflow characteristics, at which each of the aeration technologies would be the most viable option.

Looking at historical costing data of the two different aeration technologies under discussion, it is evident that the SSSA technology will have a lower initial capital cost but a higher operational cost, while the FBDA technology will have a higher initial capital cost but lower operational cost. It can therefore be safely assumed that for lower inflow rates the SSSA technology will be the most viable option and for higher inflow rates the FBDA technology will be the most viable. By comparing the total cost of the two technologies with the input of the varying inflow rates and wastewater characteristics, an inflow rate will be reached where the total cost of the two technologies are the same. For this investigation, this point will be titled the Viability Threshold Point (VTP). The VTP is important, mostly because for inflow rates lower than the VTP, the SSSA technology will be viable option and for inflow rates higher than the VTP, the FBDA technology will be.

With this information available, it will give investors and designers a benchmark as to which aeration technology should be used when and where.

2. LITERATURE REVIEW

2.1 Introduction

With electricity prices on the rise, environmental organisations driving for cleaner energy production and a struggling energy producer, energy optimisation initiatives have become the norm in South Africa. As WWTP consumes vast amounts of energy, lowering power consumption on these plants is becoming crucial. The ultimate purpose of any wastewater treatment plant is in essence to satisfy the effluent compliance requirement at the least possible cost.

As aeration forms the heart of the wastewater treatment process in terms of biological activity and effluent quality, and typically accounts for 45 to 60% of a treatment facility's total energy consumption (Roman & Mureşan, 2014), choosing the correct aeration technology is the most important choice that can be made when designing a WWTP from an energy usage perspective.

2.2 Effluent Quality

The main purpose of any WWTP is to treat wastewater so it can be safely discharged back into the environment, usually into rivers and streams. In South Africa, in order for wastewater to be safely discharged back into the environment, its constituents (COD, Nitrogen, Phosphorous, etc.) must comply with the limits as set by the Department of Water and Sanitation – General and Special Authorization (General limits or Special limits). Whether the effluent quality must comply with the General or Special Limits, is subject to the type and location of the receiving water resource. Table 1 below summarises the required limits for each constituent in the effluent, eligible for it to be safely discharged back into a water resource.

SUBSTANCE/PARAMETER	GENERAL LIMIT	SPECIAL LIMIT
Faecal Coliforms (per 100 mℓ)	1 000	0
Chemical Oxygen Demand (mg/ℓ)	75	30
pH	5.5 - 9.5	5.5 - 7.5
Ammonia (ionised and un-ionised) as Nitrogen (mg/ ℓ)	6	2
Nitrate/Nitrite as Nitrogen (mg/ ℓ)	15	1.5
Chlorine as Free Chlorine (mg/ ℓ)	0.25	0
Suspended Solids (mg/ ℓ)	25	10
Electrical Conductivity (mS/m)	70 mS/m above intake to a maximum of 150 mS/m	50 mS/m above intake to a maximum of 100 mS/m

Ortho-Phosphate as phosphorous (mg/ ℓ)	10	1 (median) and 2.5 (maximum)
Fluoride (mg/ ℓ)	1	1
Soap, oil or grease (mg/ ℓ)	2.5	0
Dissolved Arsenic (mg/ ℓ)	0.02	0.01
Dissolved Cadmium (mg/ ℓ)	0.005	0.001
Dissolved Chromium (VI) (mg/ ℓ)	0.05	0.02
Dissolved Copper (mg/ ℓ)	0.01	0.002
Dissolved Cyanide (mg/ ℓ)	0.02	0.01
Dissolved Iron (mg/ ℓ)	0.3	0.3
Dissolved Lead (mg/ ℓ)	0.01	0.006
Dissolved Manganese (mg/ ℓ)	0.1	0.1
Mercury and its compounds (mg/ ℓ)	0.005	0.001
Dissolved Selenium (mg/ ℓ)	0.02	0.02
Dissolved Zinc (mg/ ℓ)	0.1	0.04
Boron (mg/ ℓ)	1	0.5

Table 1: General and Special Limit Values. (Deprtment of Water and Sanitation, 1999)

In order to ensure the effluent from the WWTP meet the required limits, it is important to know the composition of the inflow wastewater received by the plant in order to determine the constituents that must be removed.

2.3 Wastewater Characterisation

The first step in the design of any WWTP is to characterise the inflow wastewater into its different constituents. The activated sludge system design will be only as good as the selected wastewater characteristics are representative of the particular wastewater (Wentzel & Ekama, 2003). Table 2 below shows the physical, chemical and biological transformation of the organic and inorganic constituents that take place in the biological reactor.

<table>
<tr>
<th colspan="4">WASTEWATER CONSTITUENTS</th>
<th colspan="2">REACTION</th>
<th colspan="3">SLUDGE CONSTITUENTS</th>
</tr>
<tr>
<td rowspan="6">ORGANIC</td>
<td rowspan="2">SOLUBLE</td>
<td rowspan="2">DISSOLVED</td>
<td>UNBIODE-GRADABLE</td>
<td colspan="2">NONE</td>
<td colspan="3">ESCAPES WITH EFFLUENT</td>
</tr>
<tr>
<td>BIODEGRADABLE</td>
<td colspan="2">TRANSFORMS TO ACTIVE ORGANISMS</td>
<td rowspan="9">TOTAL SETTLEABLE SOLIDS (TSS)</td>
<td rowspan="5">ORGANIC VOLATILE SETTLEABLE SOLIDS (VSS)</td>
<td rowspan="5">BIOMASS IN REACTOR ALL SETTLEABLE AND NONE SUSPENDED</td>
</tr>
<tr>
<td rowspan="4">PARTICULATE</td>
<td rowspan="2">SUSPENDED</td>
<td>UNBIODE-GRADABLE</td>
<td colspan="2">ENMESHED WITH SLUDGE MASS</td>
</tr>
<tr>
<td>BIODEGRADABLE</td>
<td colspan="2">TRANSFORMS TO ACTIVE ORGANISMS</td>
</tr>
<tr>
<td rowspan="2">SETTLEABLE</td>
<td>UNBIODE-GRADABLE</td>
<td colspan="2">ENMESHED WITH SLUDGE MASS</td>
</tr>
<tr>
<td>BIODEGRADABLE</td>
<td colspan="2">TRANSFORMS TO ACTIVE ORGANISMS</td>
</tr>
<tr>
<td rowspan="6">INORGANIC</td>
<td rowspan="2">PARTICULATE</td>
<td>SETTLEABLE</td>
<td colspan="2" rowspan="2">ENMESHED WITH SLUDGE MASS</td>
<td rowspan="4">INORGANIC SETTLEABLE SOLIDS (ISS)</td>
<td rowspan="4">INORGANIC MASS ALL SETTLEABLE AND NONE SUSPENDED</td>
</tr>
<tr>
<td>SUSPENDED</td>
</tr>
<tr>
<td rowspan="4">SOLUBLE</td>
<td>PRECIPITABLE</td>
<td colspan="2">TRANSFORMS TO SET. SOLIDS</td>
</tr>
<tr>
<td rowspan="2">BIOLOGICALLY UTILIZABLE</td>
<td rowspan="2">TRANSFER ED TO</td>
<td>SOLIDS</td>
</tr>
<tr>
<td>GAS</td>
<td colspan="3">ESCAPES AS GAS</td>
</tr>
<tr>
<td>NON PRECIPITABLE & BIOLOGICALLY UTILIZABLE</td>
<td colspan="5">ESCAPES WITH EFFLUENT</td>
</tr>
</table>

Table 2: Physical, Chemical and Biological Transformation of Organic and Inorganic Constitutes (Wentzel & Ekama, 2003)

With the composition of the inflow wastewater known and the limits of the effluent set, the correct process configuration can be chosen for the WWTP.

2.4 Process Configurations

From section 2.2 and 2.3 above it is evident that treatment of wastewater will require some sort of nitrogen or phosphorous removal or both, in order to meet the discharge limit values. The main reason behind removal of these elements is to inhibit eutrophication which has a negative impact on the natural water bodies like dams and rivers. The introduction of nutrient removal treatment systems in WWTPs are therefore very important.

2.4.1 Biological Nitrogen Removal Systems

Biological Nitrogen Removal is facilitated by organisms known as Autotrophic Nitrifier Organisms (ANOs) that has the ability to convert ammonium (NH_4^+) to nitrate (NO_3^-) with oxygen (O_2) serving as the electron acceptor, a process known as nitrification. The two basic stoichiometric redox reactions in nitrification are:

$$NH_4^+ + {}^3/_2\,O_2 \text{ (ANOs)} \rightarrow NO_2^- + H_2O + 2H^+ \tag{2-1}$$

$$NO_2 + {}^1/_2\,O_2 \text{ (NNOs)} \rightarrow NO_3^- \tag{2-2}$$

(Ekama & Wentzel, 2008)

Oxygen is predominately introduced to the wastewater by means mechanical aeration equipment. The ANOs grow in the Aerobic Zone together with the Ordinary Heterotrophic Organisms (OHOs).

ANOs are obligate aerobes, so nitrification will not happen in unaerated zones. By introducing unaerated zones into the system, the electron acceptor (O_2) is inherently removed from the process. Facultative organisms can however still survive in the unaerated zones should sufficient nitrate be available to serve as electron acceptors and sufficient COD be available as electron donors, inherently turning the unaerated zone into an anoxic zone (no dissolved oxygen, only nitrate available). In the anoxic zone, these organisms will consume nitrate (NO_3) and release nitrogen as N_2 gas to the atmosphere, a process known as denitrification. The basic stoichiometric redox reaction in denitrification is:

$$NO_3^- + 6H^+ + 5e^- \rightarrow {}^1/_2\,N_2 + 3H_2O + Energy \tag{2-3}$$

(Ekama & Wentzel, 2008)

The calculation of the nitrogen removal is essentially a reconciliation of electron acceptors (nitrate) and donors (COD) taking due account of (1) the biological kinetics of denitrification and (2) the system operating parameters (such as recycle ratios, anoxic reactor sizes) under which the denitrification is constrained to take place. (Ekama & Wentzel, 2008)

The electron donors (or COD or energy) for denitrification can come from 2 sources, (1) internal or (2) external to the activated sludge system. The former are those present in the system itself, i.e. those in the incoming wastewater or generated within the biological reactor by the activated sludge itself; the latter are organics imported to the activated sludge system and specifically dosed into the anoxic zone(s) to promote denitrification, e.g. methanol, acetate, molasses, etc. (Ekama & Wentzel, 2008) For this investigation the focus will only be on internal COD sources for denitrification.

There are many different configurations of single sludge nitrification-denitrification (ND) systems but from the point of view of the source of the organics (electron donors), these can be simplified to two basic types of denitrification or combinations of these. The two basic types utilising internal organics are (1) post-denitrification, which utilises self-generated endogenous organics and (2) pre-denitrification, which utilises influent wastewater organics. (Ekama & Wentzel, 2008). For this investigation the focus will only be on the pre-denitrification process.

2.4.1.1 Modified Ludzack-Ettinger (MLE) Process Configuration

Ludzack and Ettinger (1962) were the first to introduce a single sludge nitrification-denitrification system utilising the biodegradable organics in the influent as organics for denitrification. It consisted of two reactors in series, partially separated from each other. The influent was discharged to the first reactor or primary anoxic reactor which was maintained in an anoxic state with only mixing and no aeration. The second reactor was aerated for nitrification to be achieved. Ludzack and Ettinger reported that their system gave variable denitrification results. As a result, Barnard proposed an improvement to the Ludzack-Ettinger system by completely separating the anoxic and aerobic reactors, recycling the underflow from the Secondary Settling Tank (SST) to the primary anoxic reactor and introducing a mixed liquor recycle from the aerobic to the primary anoxic reactor (Ekama & Wentzel, 2008). This system became known as the Modified Ludzack- Ettinger process and is commonly used as the process configuration on WWTPs throughout South Africa where nitrogen removal is required.

Figure 1 below illustrates the MLE process configuration.

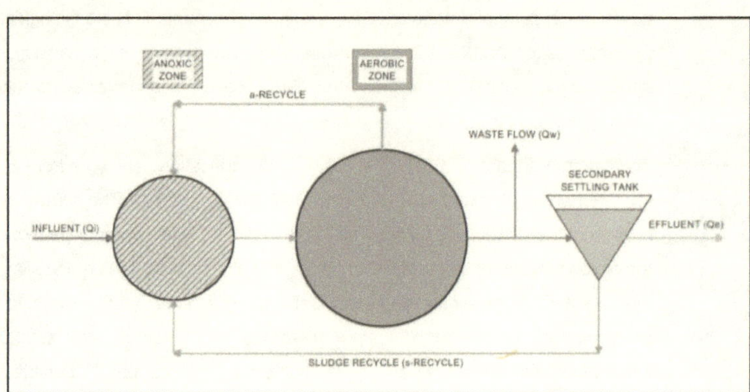

Figure 1: The MLE Process Configuration.

2.4.2 Nitrification Denitrification Biological Excess Phosphorous Removal (NDBEPR) Systems

The first indication of biological phosphate removal occurring in a wastewater treatment process was described by Srinath et al. (1959).

The development of an engineering approach to the biological phosphorous removal (BPR) process was however mainly due to work of Barnard (1974, 1975) and Nicholls (1975). They recognised that an essential prerequisite for BPR was the existence of a truly anaerobic phase, in which returned sludge and influent wastewater are mixed. The presence of an external electron acceptor in this phase limits the capacity of the BPR process. On the basis of this principle many different process configurations for biological process configuration have been proposed and constructed. (van Loosdrecht, et al., 1997)

2.4.2.1 University of Cape Town (UCT) Process Configuration

De-nitrification and biological excess phosphate removal can only take place if favourable conditions are created within the biological reactor to facilitate these processes, as is established within UCT process configuration. In the anaerobic zone, the readily biodegradable COD is taken up by a group of heterotrophic organisms known as polyphosphate accumulation organisms (PAOs), while the slowly biodegradable COD is enmeshed within the sludge mass. (Wentzel & Ekama, 1997)

Biological Excess Phosphate Removal (BEPR) is facilitated by the PAOs that can store phosphate internally as polyphosphate. When enough PAOs has been accumulated within the sludge of the activated sludge system, most of the phosphate is taken up in the anoxic and/or aerobic zones and a low effluent phosphate concentration can then be achieved. The stored phosphate is removed via the waste sludge stream from the aerobic basin. The underflow sludge from the secondary settling tanks is returned to the anoxic zone, where de-nitrification takes place. (Wentzel & Ekama, 1997).

For the UCT configuration, the nitrification/denitrification process is the same as described in section 2.4.1: Biological Nitrogen Removal Systems. The UCT process is commonly used as the process configuration on WWTPs throughout South Africa where both nitrogen and phosphate removal are required.

Figure 2 illustrates the UCT process configuration.

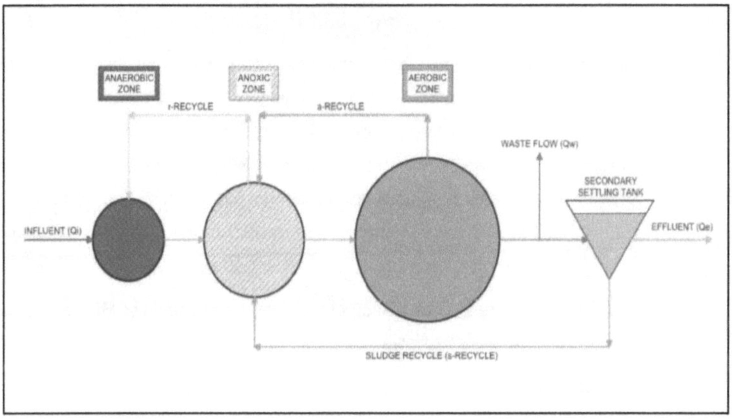

Figure 2: The UCT Process Configuration

With the correct process configuration chosen, the infrastructure associated with each configuration can be designed.

2.5 Process Infrastructure Design

2.5.1 Steady State and Dynamic Models

Steady state and dynamic models are two different methods of mathematically modelling wastewater treatment. Steady state models have constant flows and loads while dynamic models include time in their calculations and have varying loads and flows (Ekama & Wentzel, 2008) Steady state models are therefore much simpler and easy to understand and can be modelled using explicit equation on a platform like Microsoft Excel. For dynamic models very accurate inflow data is required and it is suggested that samples and measurements be taken hourly or bihourly. To gather data this accurately takes a lot of time, is very expensive and is rarely available when designing a WWTP. For this reason, steady state modelling was used for the sizing of the process infrastructure in this investigation.

2.5.2 Balanced MLE Biological Reactor Design

As explained in section 2.4.1.1, the MLE process is a single sludge nitrification - denitrification system and can therefore only facilitate biological nitrogen removal. For this investigation, the design of the MLE biological reactor was done according to the design procedure as portrayed in Chapter 5: "Nitrogen Removal" of book "Biological Wastewater Treatment: Principles, Modelling and Design" and notes presented as part of the Water Treatment courses presented at the University of Cape Town in 2016.

From this literature the calculation of the main components of MLE biological reactor can be summarised as follows:

Reactor Volume:

$$V_p = \frac{Q_{i,ADWF}S_{ti,Reactor}}{X_t}[A] \qquad\qquad (\,2\text{-}4\,)$$

Where

V_p $\qquad=\qquad$ Reactor Volume (m^3)

$Q_{i,ADWF}$ $\qquad=\qquad$ Inflow (ADWF) (m^3/d)

$S_{ti,Reactor}$ $\qquad=\qquad$ COD concentration in inflow (mgCOD/ℓ)

X_t $\qquad=\qquad$ MLSS concentration in reactor (mgTSS/ℓ)

$$A = \left(1 - f_{S'up} - f_{S'us}\right)\frac{Y_H R_s}{(1 + b_{HT}R_s)}\left(1 + f_H b_{HT}R_s + f_{i,OHO}\right) + \left[\frac{f_{S'up}}{f_{cv}} + \frac{X_{IOi}}{S_{ti}}\right]R_s$$

$f_{S'up}$	=	Fraction of total COD which is unbiodegradable particulate
$f_{S'us}$	=	Fraction of total COD which is unbiodegradable soluble
Y_H	=	Biomass Yield of OHO's
R_S	=	Sludge age (for balanced sludge calculation, see below)
b_{HT}	=	Endogenous respiration rate OHO's
f_H	=	Fraction of endogenous residue in OHO's
$f_{i,OHO}$	=	VSS/TSS ratio of activated sludge
f_{cv}	=	COD/VSS ration of sludge
X_{IOi}	=	Concentration of unbiodegradable particulate organics in inflow
S_{ti}	=	Total influent COD concentration

Total oxygen demand:

$$FO_t = FO_c + FO_n - FO_d \qquad (\,2\text{-}5\,)$$

Where:

FO_t	=	Flux of total oxygen demand (kgO$_2$/day)
FO_c	=	Flux of carbon oxygen demand (kgO$_2$/day)
FO_n	=	Flux of nitrification oxygen demand (kgO$_2$/day)
FO_d	=	Flux of denitrification oxygen demand (kgO$_2$/day)

$$OUR = {FO_t}/{V_{Aer}} \times 24 \qquad (\,2\text{-}6\,)$$

Where:

OUR	=	Oxygen utilisation rate mgO$_2$/(l/h)
V_{Aer}	=	Volume of aeration basin (ℓ)
24	=	24 hours

As costing of the infrastructure played such a critical part in the investigation, it was important that the optimal sizing of the infrastructure be used. In an MLE configuration there is a clearly defined optimum configuration that results in the smallest reactor for a given influent flow which is referred to as a balanced MLE system. In order to facilitate a balanced MLE system, the balanced SRT (Sludge Retention Time) or sludge age must be calculated. The SRT governs the maximum unaerated fraction and in a balance MLE system, the anoxic fraction is set equal to the maximum unaerated fraction. The size of the anoxic zone, on its part sets the denitrification potential of the system. On their part, the a- and s- recycle returns set the nitrate load from the aerobic zone to the anoxic zone. With the a- and s-recycle flow known (for this investigation the a-recycle was set to 6 times the inflow and s-recycle was set equal to the inflow) the nitrate load can be determined. In a balanced MLE system, the SRT is chosen so that the denitrification potential is equal to the nitrate load. This SRT is known as the balanced SRT of the MLE system. From the notes on work done by G.A. Ekama: "Calculation procedure for balanced SRT of MLE system", the following explicit equation was used to calculate the balanced SRT.

$$R_{sBalMLE} = \frac{C.E_{MLE} + D - A.B + A.S_f K_{2T} \frac{Y_H}{\mu_{AmT}} - E_{MLE}.f_n \left[A.Y_H + S_{ti} \frac{f_{Srup}}{f_{cvUPO}} \right]}{A(B.b_{HT} + K_{2T} Y_H) - A.S_f b_{AT} K_{2T} \frac{Y_H}{\mu_{AmT}} - b_{HT}(C.E_{MLE} + D) + E_{MLE}.f_n b_{HT} \left[A.Y_H f_H + S_{ti} \frac{f_{Srup}}{f_{cvUPO}} \right]} \qquad (2\text{-}7)$$

Where,

$$A = S_{ti}\left(1 - f_{S'us} - f_{Srup}\right)$$

$$B = f_{Sb's} \frac{\left(1 - f_{cv,OHO} Y_H\right)}{2.86}$$

$$C = N_{ti} - N_{te}$$

$$D = \frac{a_{prac} O_a + s O_s}{2.86}$$

$$E_{MLE} = \frac{\left(a_{prac} + s\right)}{\left(a_{prac} + s + 1\right)}$$

With the optimum mixed liquor suspended solids (MLSS) concentration known from section 2.5.5 (see below) and the balanced SRT calculated from equation (2-7), the sizing of the MLE biological reactor was completed.

2.5.3 Balanced UCT Biological Reactor Design

The UCT process is a single sludge NDBEPR system and is capable of biological removal of nitrogen and phosphorous. The design of the UCT biological reactor was done according to the design procedure as portrayed in the paper by M.C. Wentzel and G.A. Ekama "Principles in the design of single-sludge activated-sludge systems for biological removal of carbon, nitrogen and phosphorus" (1997) and notes presented as part of the Water Treatment courses presented at the University of Cape Town in 2016.

From this literature the calculation of the main components of UCT biological reactor can be summarised as follows:

Reactor Volume:

$$V_p = \frac{Q_{i,ADWF} S_{ti,Reactor}}{X_t} [A] \qquad\qquad (\ 2\text{-}8\ )$$

Where

$Q_{i,ADWF}$	=	Inflow (ADWF) (m^3/d)
$S_{ti,Reactor}$	=	COD concentration in inflow (mgCOD/ℓ)
X_t	=	MLSS concentration in reactor (mgTSS/ℓ)

$$A = \left(1 - f_{S'up} - f_{S'us}\right)\left[\left(1 - \frac{\%}{100} f_{Sb's}\right)\frac{Y_H R_s}{(1 + b_{HT} R_s)}\left(1 + f_H b_{HT} R_s + f_{i,OHO}\right)\right.$$

$$+ \left(\frac{\%}{100} f_{Sb's}\right)\frac{Y_G R_s}{(1 + b_{GT} R_s)}\left(1 + f_{EG} b_{GT} R_s + 3.286(f_{XBGP} - f_{XBGPBM})\right.$$

$$\left.\left. + f_{iPAOBM}\right)\right] + \left[\frac{f_{S'up}}{f_{cv}} + \frac{X_{IOi}}{S_{ti}}\right] R_s$$

$f_{S'up}$	=	Fraction of total COD which is unbiodegradable particulate
$f_{S'us}$	=	Fraction of total COD which is unbiodegradable soluble
$f_{Sb's}$	=	Fraction of biodegradable COD which is readily biodegradable (soluble)
Y_H	=	Biomass Yield of OHO's
R_S	=	Sludge age (for balanced sludge calculation, see below)
b_{HT}	=	Endogenous respiration rate OHO's
f_H	=	Fraction of endogenous residue in OHO's
$f_{i,OHO}$	=	VSS/TSS ratio of activated sludge
Y_G	=	Biomass Yield of PAO's
b_{GT}	=	Endogenous respiration rate PAO's
f_{EG}	=	Fraction of endogenous residue in PAO's
f_{XBGP}	=	Total PAO phosphorus content
f_{XBGPBM}	=	Phosphorus content of PAO biomass
f_{iPAOBM}	=	ISS content of PAO biomass

f_{cv} = COD/VSS ration of sludge

X_{IOi} = Concentration of unbiodegradable particulate organics in inflow

S_{ti} = Total influent COD concentration

Total oxygen demand:

Calculation of the total oxygen is similar to the calculation used for the MLE system.

The same as for the MLE process, the UCT process also has an optimum configuration giving the smallest reactor for a said inflow, this is referred to as a balanced UCT system. The UCT configuration, however, has added complexity as the activated sludge contains both OHOs and PAOs. Like for an MLE system, a balanced sludge age for the UCT system ensures that the denitrification potential of the anoxic reactor is equal to the equivalent nitrate load of the recycles (for this investigation the a-recycle was set to 6 times the inflow, r-recycle set equal to the inflow and s-recycle also set equal to the inflow) to the anoxic zone. For the UCT balanced SRT no explicit equation has been formulated and an iterative process must be followed to calculated the balanced SRT.

Notes on work done by G.A. Ekama: "Calculation procedure for balanced SRT of UCT system" gives the following equations:

$$LHS = D_{p1} = A\left[B\left(1 - \frac{\%}{100}\right) + (1 - F)K'_{2T}\left\{1 - \frac{S_f\left(b_{AT} + \frac{1}{R_s}\right)}{\mu_{AMT}} - f_{xa}\right\}\frac{Y_H R_s}{1 + b_{HT}R_s}\right] \qquad (\,2\text{-}9\,)$$

Where:

$$A = S_{ti}(1 - f_{S'us} - f_{S'up})$$

$$B = f_{Sb's}\frac{1 - f_{cv,OHO}Y_H}{2.86}$$

$$F = \frac{\%}{100}f_{Sb's}$$

$$RHS = D_{p1} = \frac{a_{prac}O_a + sO_s}{2.86} + \frac{(a_{prac} + s)}{(a_{prac} + s + 1)}(C - N_s) = D_{UCT} + E_{UCT}G_{UCT} \qquad (\,2\text{-}10\,)$$

Where:

$$D_{UCT} = \frac{a_{prac}O_a + sO_s}{2.86}$$

$$E_{UCT} = \frac{a_{prac} + s}{a_{prac} + s + 1}$$

$$C = N_{ti} - \frac{K_{nT}}{S_f - 1} - N_{ouse}$$

$$G_{UCT} = C - f_n \left\{ A \left[(1-F)\frac{Y_H(1 + f_H b_{HT}R_S)}{(1 + b_{HT}R_S)} + F\frac{Y_G(1 + f_G b_{GT}R_S)}{(1 + b_{HT}R_S)} \right] + \frac{f_{S'up}}{f_{cv,UPO}} S_{ti} \right\}$$

For the calculation of the balanced SRT, the SRT (R_s in the equations) must be chosen as part of an iterative process until LHS = RHS. With the optimum MLSS concentration known from section 2.5.5 (see below) and the balanced SRT calculated from the iterative process, the sizing of the UCT biological reactor was completed.

2.5.4 Secondary Settling Tank Design

As can be seen from Figure 1 and Figure 2 in section 2.4, the SSTs follow the biological reactors. In the SSTs physical separation of the solids from the treated effluent occurs and the biomass (underflow) is returned to the biological reactor via the Return Activated Sludge (RAS) flow (s-recycle). It is imperative that the separation process happen thoroughly and that the correct underflow rate is chosen to ensure as little as possible sludge carryover to the final effluent occurs.

The design of the SSTs was done according to the One-Dimensional Flux Theory (1DFT) as described by G.A. Ekama in the paper "Estimation of Secondary Settling Tank Capacity with the One-Dimensional Flux Theory". From this literature the calculation main components of Secondary Settling Tank can be summarised as follows:

Flux constants V_0 and n:

$$SSVI = DSVI \times 0.67 \tag{2-11}$$

Where:

$SSVI$ =   Stirred sludge volume index (mℓ/g)

$DSVI$ =   Diluted sludge volume index (mℓ/g)

$$V_0/_n = 67.9\, e^{-0.016(SSVI)} \tag{2-12}$$

Where:

V_0 = flux constant (m/d)

n = flux constant (m³/kg)

$$n = 0.88 - 0.393 \log\left(V_0/n\right)$$

(2-13)

Maximum up-flow-velocity:

$$q_{Amax} = V_0\, e^{-nXt}$$

(2-14)

Where:

q_{Amax} = maximum up-flow-velocity (m/h)

X_t = TSS concentration in reactor (kgTSS/m³)

Calculating SST area:

$$A_{SST} = PWWF/q_{Amax}$$

(2-15)

Where:

A_{SST} = Area of SST (m²)

$PWWF$ = Peak wet weather flow (m³/h

2.5.5 Reactor/SST Cost Optimisation

As the biological reactor and SST forms the basis of the process configurations and the costing played a critical part in the investigation, it was important that the cost of this infrastructure be kept to a minimum. In order to optimise costing of the Reactor-SST system, the optimum MLSS concentration was calculated based on the notes presented as part of the Water Treatment courses presented at the University of Cape Town in 2016 "Optimising Activated Sludge Systems":

Biological Reactor

From the steady state Reactor model, equation (2-16) can be generated to calculate the reactor volume (V_p).

$$V_p = \frac{Q_{i,ADWF}\, S_{ti,Reactor}}{X_t}\,[A]$$

(2-16)

Where

$Q_{i,ADWF}$ =   Inflow (ADWF) (m³/d)

$S_{ti,Reactor}$ =   COD concentration in inflow (mgCOD/ℓ)

X_t =   MLSS concentration in reactor (mgTSS/ℓ)

$$A = \left(1 - f_{S'up} - f_{S'us}\right)\frac{Y_H R_s}{(1 + b_{HT}R_s)}\left(1 + f_H b_{HT}R_s + f_{i,OHO}\right) + \left[\frac{f_{S'up}}{f_{cv}} + \frac{X_{IOi}}{S_{ti}}\right]R_s$$

for N removal systems (MLE configuration in this investigation) and

$$A = \left(1 - f_{S'up} - f_{S'us}\right)\left[\left(1 - \frac{\%}{100}f_{Sb's}\right)\frac{Y_H R_s}{(1 + b_{HT}R_s)}\left(1 + f_H b_{HT}R_s + f_{i,OHO}\right)\right.$$

$$+ \left(\frac{\%}{100}f_{Sb's}\right)\frac{Y_G R_s}{(1 + b_{GT}R_s)}\left(1 + f_{EG}b_{GT}R_s + 3.286(f_{XBGP} - f_{XBGPBM})\right.$$

$$\left.\left. + f_{iPAOBM}\right)\right] + \left[\frac{f_{S'up}}{f_{cv}} + \frac{X_{IOi}}{S_{ti}}\right]R_s$$

for N & P removal systems (UCT configuration in this investigation).

From equation (2-16) it is evident that the reactor volume will decrease with an increase in MLSS concentration (X_t).

Secondary Settling Tank

From the 1DFT, equation (2-17) can be generated to calculate the SST Area (A_{SST}).

$$A_{SST} = \frac{1000 f_q Q_{i,ADWF} / 24}{0.8 V_0 exp(-nX_t)} \tag{2-17}$$

Where

f_q =   PWWF / ADWF ratio

$Q_{i,ADWF}$ =   Inflow (ADWF) (m³/d)

V_0 =   flux constant (m/d)

n =   flux constant (m³/kg)

X_t =   MLSS concentration in reactor (mgTSS/ℓ)

0.8   =   Flux rating.

From equation (2-17) it is evident that SST area will increase with an increase in MLSS concentration (X_t).

The size of the reactor and SST are therefore linked by the MLSS concentration. The optimum MLSS concentration is the concentration at which the combined cost of the reactor and SST is a minimum. This optimum concentration was calculated using the above equations and cost functions for the reactor and SST (see section 2.8 below). With the optimum MLSS concentration known and from the equation presented in section 2.5.2, 2.5.3 and 2.5.4, the design of the Reactors and SSTs were completed.

2.5.6 Blower House Building

The purpose of the blower house building is to house the blowers and associated equipment, which normally include the electrical equipment such as the motor control centers (MCCs), transformers and switchgear.

For the blower room itself special care must be given to include noise attenuation, coarse air filtration, high level extraction fans and a gantry assembly for installation and removal of the blowers.

Blower houses range in size and type depending on type and size of blower equipment to be installed as well as geographical location of the WWTP. The size of the blower house building for this investigation will therefore be based on the size and type of the blower equipment installed. The figure below gives the standard sizes of the blower equipment to be installed.

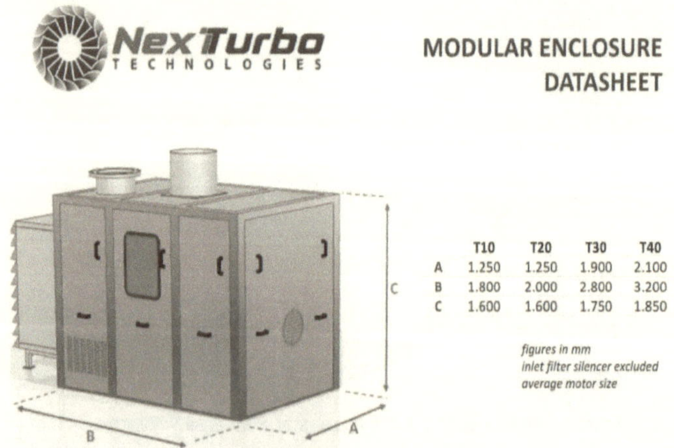

	T10	T20	T30	T40
A	1.250	1.250	1.900	2.100
B	1.800	2.000	2.800	3.200
C	1.600	1.600	1.750	1.850

Figure 3: Modular Enclosure Datasheet of Blowers

In order to do proper maintenance on the blowers, it is common practice to allow for an area of at least 2m on all sides of the blowers. From previous projects it is also assumed that the area required to house the electrical equipment will be a similar size to that of the area required for the blowers. The table below gives the estimated areas required.

Blower Type	Number of Units to be installed	Area required for Blowers	Area of Blower Building
GT-B-T10	2 (1 duty, one standby)	2x (1.25+2+2) x (1.80+2+2) = 60.90m^2	2 x 60.90 = ± 120 m^2
GT-B-T20	2 (1 duty, one standby)	2x (1.25+2+2) x (2.00+2+2) = 63.00m^2	2 x 63.00 = ± 130 m^2
GT-B-T30	3 (2 duty, one standby)	3x (1.90+2+2) x (2.80+2+2) = 120.36m^2	2 x 120.36 = ± 240 m^2
GT-B-T40	3 (2 duty, one standby)	3x (2.10+2+2) x (3.20+2+2) = 131.76m^2	2 x 131.76 = ± 260 m^2

Table 3: Blower Building Sizing

For more information on the blower equipment, see section 2.8.1.6.

2.6 Aeration on WWTPs

2.6.1 Theory of Oxygen Transfer

As described in section 2.4, oxygen forms an essential part for the biological treatment of wastewater. Several methods exist to introduce oxygen into water, but basically all of them rely on the process of diffusion through an air/liquid interface and then dispersion throughout the liquid body by diffusion and convection. (Bratby, 1982)

Oxygen transfer across the interface occurs in three stages:

1) Oxygen molecules from gas (air or oxygen) are initially transferred to the air/liquid interface until a stage is reached where gas molecules dissolving into the liquid and gas molecules escaping out of solution reach an equilibrium, thus the layer becomes saturated with oxygen;

2) Oxygen molecules dissolved to saturation at the interface pass through the interfacial film by molecular diffusion. The interfacial film is composed of water molecules which, because of the air phase on the other side of the film, are orientated with their negative oxygen ends facing the interface (Horne, 1969);

3) Oxygen is distributed throughout the body of liquid by diffusion and convection.

(Bratby, 1982)

2.6.2 Oxygen Transfer rate

Oxygen transfer rate is defined as the mass of oxygen an aeration device can introduce into a body of water per unit of time and unit of power absorbed. This rate can be represented as the following equation:

$$R_{std} = {K_{La} C_s V_{Aer}} / {kW} \qquad\qquad (\text{2-18})$$

where

R_{std} - Oxygen transfer rate at standard conditions (kgO/kWh);

K_{La} - Mass transfer coefficient;

C_S - Saturation concentration of oxygen in liquid (mg/ℓ);

V_{Aer} - Volume of liquid in aeration reactor (m³);

kW - Power absorbed.

(Bratby, 1982)

The figure for R_{std} is quoted by the manufacturer and will vary according to the specific equipment under discussion. These R_{std} figures quoted by the manufacturer are however only valid under standard conditions. The following effects must be included to allow for non-standard (site) conditions:

2.6.2.1 Effect of impurities on k_{La}

As the mass transfer coefficient (K_{La}) is based on clean tap water, the effect that impurities in the water has on K_{La} must be calculated. The factor "α" is incorporated into the equation to allow for this effect. (Bratby, 1982). For accurate determination of α values it is suggested that laboratory tests be performed.

2.6.2.2 Effect of impurities on C_s

Impurities will not only effect K_{La} as described in section 2.6.2.1 but also the saturation concentration of the oxygen into the liquid (C_s). The correction factor "β" is used to correct the test system oxygen transfer rate for differences in oxygen solubility. Values of β vary from about 0.7 to 0.98. A β value of 0.95 is commonly used for wastewater. (Metcalf & Eddy, 2004)

2.6.2.3 Effect of atmospheric pressure on C_s

Pressure for Standard Conditions are set at 1 atmosphere (760 mm Hg). For pressures other than that, the saturation concentration in water C_s, is given by:

$$C_s = C_{S_{std}}\left(\frac{P_{site} - p_{site}}{P_{std} - p_{std}}\right)$$
(2-19)

where

C_S	-	saturation concentration of oxygen at site;
C_{Sstd}	-	saturation concentration of oxygen at 1 atmosphere;
P_{site}	-	Pressure (mm Hg) at site;
P_{std}	-	1 atmosphere (760mm Hg);
p_{std}	-	saturated water vapour pressure at standard temperature;
p_{site}	-	saturated water vapour pressure at particular water temperature;

(Bratby, 1982)

Adjusting pressure (P) for increase in attitude:

$$P = 10^{2.88117 - 0.000053407 \times Alt)}$$
(2-20)

Adjusting Saturated Water Vapour Pressure for increase in temperature:

$$p = 17.51(1.0639)^{(T-20)}$$ (2-21)

2.6.2.4 Effect of temperature on k_{La} and C_s

The temperature effect on K_{La} is defined by the following relationship (Eckenfeller and Ford, 1968):

$$(K_{La})_T = (K_{La})_{20°C}\theta^{(T-20)}$$ (2-22)

The temperature effect on C_s is given by the following equation:

$$(C_s)_T = (C_s)_{20°C}\,{51.6}/{31.6+T}$$ (2-23)

Typical θ values are in the range of 1.015 to 1.040. The θ of 1.024 is typical for both diffused and mechanical aeration devices. (Metcalf & Eddy, 2004)

2.6.2.5 Overall correction to oxygen transfer rate

The above effects can be incorporated into one overall correction factor that must be applied to the manufacturers quoted R_{std} figure, given by the following:

$$\frac{R_{act}}{R_{std}} = \alpha\theta^{(T-20)}\frac{\left[\left(\frac{P_{site}-p_{site}}{P_{site}-p_{std}}\right)\left(\frac{51.6}{31.6+T}\right)\beta C_s - C_L\right]}{C_s}$$ (2-24)

2.6.3 Oxygen Content of Air

As most aeration devices introduces air into the wastewater but only oxygen is used for the biological process, the oxygen content of air must be calculated.

The mole fraction of air (at sea level) according the book Atkin's Physical Chemistry (1982) is given as:

N_2	=	2.7 moles
O_2	=	0.72 moles
Ar	=	0.032 moles
CO_2	=	0.001 moles
Total =		**3.453 moles**

This means there is 0.72 mol O_2 in 3.453 mol air. These fractions do vary slightly with altitude, as the heavier gases decrease with altitude. The variation is however so miniscule that it has no significant effect on the calculations and can therefore be ignored.

The molar fraction of air is therefore calculated as:

$$Molar\ Fraction\ of\ O_2 = {0.72\ mol\ O_2}/{3.453\ mol\ air} = 0.2085\ {mol\ O_2}/{mol\ air} \qquad (\ 2\text{-}25\ )$$

giving 0.2085 x 32g (molar mass of O_2) = 6.672g O_2 in 1 mol of air.

From the perfect gas law, the volume of air is calculated:

$$PV = nRT \qquad\qquad\qquad (\ 2\text{-}26\ )$$

where:

P	=	pressure (N/m², where 1 kPa = 10^3 N/m²);
V	=	Volume (m³);
n	=	no. moles;
R	=	Universal gas constant (8.31441 J/(mol.K));
T	=	Temperature (K).

Rearranging equation (2-26) the volume of air at standard temperature and pressure, STP (273.15 K and 1.01325 x 10^5 N/m²), can be calculated.

$$V/n = RT/P \qquad\qquad\qquad (\ 2\text{-}27\ )$$

Solving equation (2-27) gives 1 mol of air occupies 0.0224 m³ @ STP. As 1 mol of air contains 6.672g O_2, there is 0.298 kgO in 1Nm³ air @ STP

Combing the above equations into a single equation, the O_2 content of air (Con_{O2}) is given by:

$$Con_{O2} = \frac{\eta_{O2} MW_{O2} P}{RT\,1{,}000} \qquad\qquad (\ 2\text{-}28\ )$$

Where:

Con_{O2}	=	oxygen content of air (kgO/m³);
η_{O2}	=	mole fraction of O_2 in air;
	=	0.2085 (mol/mol);
MW_{O2}	=	molecular weight of O_2;
	=	32 (g/mol);
1000	=	conversion from g to kg.

Solving equation (2-28) gives:

Con_{O2} = 0.298 kgO/m³ @ 273.15 K and standard pressure of 1.01325 x 10^5 N/m².

(Wagner & Pöpel, 1998)

2.6.4 Peak Flow Factors

The size of a WWTP is normally measured in Mℓ/d, which refers to the Average Dry Weather Flow (ADWF) the plant can treat. The actual flow to the WWTP can however vary considerably throughout the day (see Figure 4 below) and is dependent on various factors such as size of catchment area, land use (industrial residential or commercial), type of sewage system (gravity vs pumping), ect. These flow variations must be considered when designing a WWTP and for this reason a peaking factor known as the Peak Flow Factor must be incorporated. Various methods exist to calculate this factor but the one most commonly used in South Africa is the Harmon Factor (see equation (2-29)). By incorporating this factor, the Peak Dry Weather Flow (PDWF) can be calculated.

$$PFF = 1 + \frac{14}{4 + \sqrt{POP}}$$

(2-29)

Where:

PFF =   Peak Flow Factor

POP =   Population in thousands

(van Vuuren & van Dijk, 2011)

Together with these flow variations over the day, the flow variations due to infiltration into the sewage system (normally associated with precipitation and rising groundwater) also plays a role. Again, many factors will influence the amount of infiltration, such as the amount of precipitation, age of sewage system, type of sewage system, ect. The infiltration ratio is normally measured as a percentage of the PDWF and can vary considerably (normally 15% - 50%). The major influence on the flow to WWTP will be when the PDWF and peak infiltration happens simultaneously. This flow is known as the Peak Wet Weather Flow (PWWF).

This PWWF influence the sizing of various components of the WWTP, but relevant to this investigation, especially that of the SSTs and aeration requirements.

2.6.5 Effect of Cycle Flow and Load Conditions on Oxygen Demand

The flow and load to a WWTP may vary considerably throughout the day and the peak flow and pollutant load can be several times higher than the average. For such cyclic flow and load conditions, the mass of sludge produced and plant volume remain unchanged from the steady state values, but the oxygen demand fluctuates in response to the varying load. The peak daily oxygen demand can therefore be considerably higher than the average daily oxygen demand and needs to be accurately determined at design stage in order to provide sufficient aeration capacity during peak oxygen demand periods. (Musvoto, et al., 2002)

The graph below shows typical cycle flow conditions and how it various to ADWF over the day:

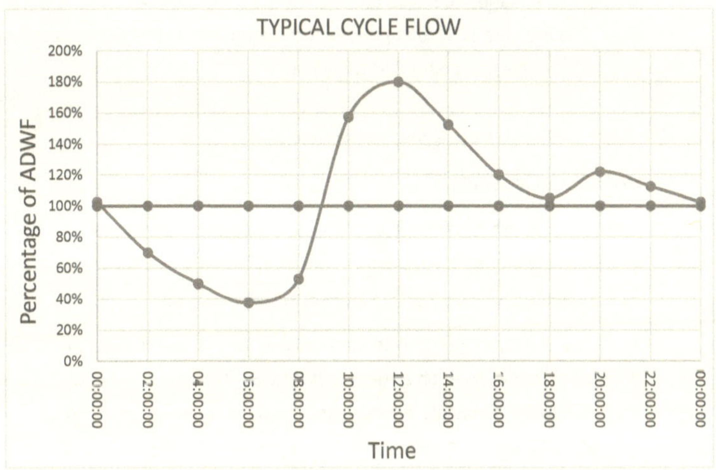

Figure 4: Typical Cycle Flow over a Day

There are a number of methods which designers can use to determine the peak oxygen demand depending on the data available to them. If diurnal load variation data is available, the method developed by Ekama and Marais (1978) for estimating peak oxygen demand in fully aerobic activated sludge systems can be used. This is done empirically from the finding that the amplitude (Peak/Average - 1) of the Oxygen Utilization Rate (OUR) under cyclic loading conditions is half that of the total oxygen demand (TOD) load wave, i.e.

$$a_0 = d_{peak} a_L \qquad\qquad\qquad (2\text{-}30)$$

Where:

a_0 =   amplitude of the OUR response wave = (Peak OUR/Average OUR) – 1;

d_{peak} =   the damping of the amplitude of peak OUR relative to the amplitude of the peak TOD load

a_L =   amplitude of the TOD load wave = (Peak TOD/Average TOD) – 1

The average OUR is the oxygen utilisation rate including nitrification and denitrification calculated from the steady state activated sludge model. The Peak and Average TOD are determined from the cyclic flow and load variation as follows: The COD load at any time in the day is the product of the flow and COD concentration at that particular time. Similarly, for the TKN load. Now the TOD load at each time is the COD load plus 4.57 times the TKN load, and the Peak TOD is the maximum TOD value of the day. The average TOD load per day is the area under the TOD load versus time of day plot. Hence, the peak OUR is given by:

$$Peak\ OUR = \left(1 + d_{peak} \times a_L\right) \times average\ OUR \qquad\qquad (\ 2\text{-}31\ )$$

(Musvoto, et al., 2002)

From Ekama and Marais (1978), d_{peak} was calculated as approximately 0.5 for fully aerobic activated sludge systems. From Musvoto, et al. (2002), d_{peak} was calculated as 0.28 at 20°C and 0.32 at 22°C for ND-BEPR systems. (Musvoto, et al., 2002)

As diurnal flow information was not available for this investigation, the peak oxygen demand was calculated on a safety factor basis. This safety factor will be site specific and will depend on factors like factors such as the layout of the sewerage system, the type of and intensity of industries, etc. Metcalf and Eddy (1991) propose a safety factor of twice the average biological oxygen demand load is used to design the aeration equipment. This is found to be extremely high if only domestic wastewater is treated. From previous experience on WWTP treating domestic wastewater, the peak factor was found to be ± 1.25 the average oxygen requirement. This safety factor (1.25 the average biological oxygen demand load) was therefore used in this investigation.

2.6.6 Slow Speed Surface Aeration (SSSA)

The introduction of air into the mixed liquor by surface aerators is based on violent surface agitation of the water to form small droplets in order to diffuse oxygen into the water. For the introduction of air by SSSA, surface aerators with a maximum rotational speed of 45 rpm was used in this investigation.

As described in section 2.6.2, oxygen transfer rate (R_{std}) is manufacturer specific. For the modeling purposes of this investigation an average R_{std} figure was used based on equipment commonly used in the South African market.

Taking wastewater characteristics and site-specific conditions (as described in section 2.6.2) into account, R_{act} is given by the following equation:

$$R_{act} = \alpha\theta^{(T-20)} \frac{\left[\left(\frac{P-p}{760-p_{std}}\right)\left(\frac{51.6}{31.6+T}\right)\beta C_s - C_L\right]}{C_s} \times R_{std} \qquad (2\text{-}32)$$

For this investigation the following values were used:

> α =   0.80;

> β =   0.95;

> θ =   1.024;

> R_{std} =   2.0.

With R_{act} calculated, the oxygen utilisation rate (OUR) and volume of the reactor known from the biological reactor design, the total power requirement for the plant was calculated.

$$P_w = R_{act} \times OUR \times V_{Aer} \qquad (2\text{-}33)$$

Where:

P_w =   Total power requirement (kW);

OUR =   Oxygen utilisation rate (mgO/(l.h));

V_{Aer} =   Volume of aeration basin (m^3).

With the total power requirement known, the correct size aerators can be chosen. As aerators are only available in standard sizes (see section 2.8.1.5 below), the sizes of the aerators were chosen to best suit the power requirement calculated above.

2.6.7 Fine Bubble Diffused Aeration

The typical configuration of a diffused aeration system involves an air compressor (blower) that supplies air through tubing to diffusers located at the bottom of the reactor basin (aeration zones). When the air reaches the diffuser, the diffuser breaks it up into tiny bubbles that are released into the mixed liquor. The bubbling action of the air rising to the surface causes both mixing and aeration of the mixed liquor.

From the work done by Wagener and Pöpel (1998), the method for sizing diffused aeration equipment are as follow and were used in this investigation:

➢ As explained in section 2.6.3 the air supplied by the blowers has a specific oxygen content. The oxygen content is dependent on the temperature and air pressure of the specific WWTP site and must be taken into account.

➢ The oxygen in the air supplied is transferred to the water via the diffusers which have a specific transfer efficiency, which is specified by the manufacturer/supplier, and is dependent on water depth, diffuser density and airflow rate.

➢ The oxygen transferred to the water must always be able to meet the biological demand. The aeration equipment must therefore be sized to meet maximum oxygen demand.

➢ As with the SSSA system, the specified oxygen transfer efficiency is for "standard" conditions, whereas the oxygen demand is at "site" conditions. An adjustment must therefore be made to convert "standard" conditions to "site" conditions. To achieve this, the oxygen transfer rate (OTR) must be adjusted from "standard" conditions to "site" conditions.

➢ With the oxygen demand, oxygen transfer rate and oxygen content of air known, the required airflow rate can be determined.

➢ Based on the airflow rate, the number of diffusers can be selected together with the diffuser density. The diffuser density (DD) should not exceed 25%.

➢ With the airflow rate known, the air pipe sizes can be calculated.

➢ From the airflow rates and pipe sizes, the headloss through the pipes and diffusers can be determined.

➢ The total headloss is the sum of the above headlosses, the static head created by the depth of immersion of the diffusers and the fouling encountered at the diffusers.

➢ The blowers are sized based on the airflow rate and headloss to overcome (pressure).

(Wagner & Pöpel, 1998)

2.6.7.1 Fouling Factor

Fouling of the diffuser system can be split into two categories i.e. internal fouling and external fouling. Internal fouling is caused by impurities in the compressed air, whereas external fouling is caused by the formation of biological slimes and inorganic precipitants (Metcalf & Eddy, 2004). This fouling has an effect on the diffuser mass transfer and can materially alter the air volume required and thus power required. The correction factor "F" is used to correct the test system oxygen transfer rate for the fouling on the diffuser system. Values of F vary from about 0.65 to 0.90 depended on cleaning regimes and type of diffuser system.

2.6.7.2 Airflow Calculation

Applying a similar approach as was used with the SSSA (oxygen transfer rate (R_{act})) calculation (section 3.3.1), the oxygen transfer rate (OTR) for the diffused aeration is converted from standard conditions to site conditions with the following equation:

$$\frac{OTR_{site}}{OTR_{std}} = \left\{ \alpha\theta^{T-20} \frac{\left[\left(\frac{P_{site} - p_{site}}{P_{std} - p_{std}} \right) \left(\frac{P_{site} + 73.53hf - P_{site}}{P_{site} - p_{site}} \right) \left(\frac{51.6}{31.6 + T} \right) \beta C_{Sstd}F - C_L \right]}{C_s} \right\} = \{A\} \qquad (\,2\text{-}34\,)$$

Where:

OTR_{site}	$=$	Oxygen transfer rate (kgO/h) for site conditions;
OTR_{std}	$=$	Oxygen transfer rate (kgO/h) for standard conditions (20°C, 101.325 kPa);
α	$=$	(KLa of mixed liquor)/(KLa clean tap water)
	$=$	0.5 (for this study);
θ^{T-20}	$=$	effect of temperature on KLa;
θ	$=$	1.024 (for this study);
P_{site}	$=$	atmospheric pressure on site (mm Hg);
p_{site}	$=$	saturated vapour pressure at site, for site temperature (mm Hg);
P_{std}	$=$	standard pressure;
p_{std}	$=$	standard saturated vapour pressure;
73.53	$=$	conversion factor for m water to mm Hg;
h	$=$	submersion depth for diffusers (m);
f	$=$	fraction of submerged depth (from surface) at which pressure corresponds to the average saturation concentration
	$=$	0.325 (for this study);
T	$=$	temperature in °C;

β = effect of impurities on CS (CS mixed liquor / CS of clean tap water)

= 0.95 (for this study);

C_{Sstd} = saturated concentration of DO under standard conditions (mgO/l)

= 9.07;

F = fouling factor

= 0.9 (for this study);

C_L = Residual dissolved oxygen concentration (2mg/l)

= 2.0 (for this study).

It is also known that:

$$OSR_{std} = Con_{O2,std} \times Q_{Air,std} = {OTR_{std}}/{SOTE_{std}} \qquad (\ 2\text{-}35\ )$$

Where:

OSR_{std} = Oxygen supply rate (kgO/h) under standard condition;

$Con_{O2,std}$ = oxygen content of air (kgO/m^3);

$Q_{Air,std}$ = Air flow rate (m^3/h);

$SOTE_{std}$ = Standard oxygen transfer efficiency.

$$SOTE_{std} = OTE \times submergence\ depth \qquad (\ 2\text{-}36\ )$$

and

$$OTR_{site} = AOR \qquad (\ 2\text{-}37\ )$$

therefore,

$$AOR_{site} = OTR_{std}\{A\} = Con_{O2,std} \times Q_{Air,std} \times SOTE_{std}\{A\} \qquad (\ 2\text{-}38\ )$$

Where:

AOR = Actual oxygen requirement (kgO/h);

$\{A\}$ = from equation (2-34).

With this, $Q_{Air,std}$ can be solved

$$Q_{Air,std} = {AOR_{site}}/{(Con_{O2,std} \times SOTE_{std}\{A\})} \qquad (\ 2\text{-}39\ )$$

With $Q_{Air,std}$ known, the airflow rate of the air intake to the blower at the actual temperature and pressure must be calculated.

In the calculations above, the air flow rate calculated is for the conditions within the aeration zone. In effect this would be the volume of air at the blower intake, since this air supplies the mass of oxygen required (OSR). The temperature of the air taken into the blower may however be very different from these conditions, and so the airflow rate should be calculated for the actual air temperature and pressure at the blower. To compensate for this, the actual temperature and pressure at the blower must be taken into account, giving the following equation.

(Wagner & Pöpel, 1998)

$$Q_{Air,Blower} = {AOR_{site}} \Big/ {\left(Con_{O2,Blower} \times SOTE_{std}\{A\} \right)} \qquad (\,2\text{-}40\,)$$

Where:

$Con_{O2,Blower}$ $\qquad$ = $\qquad$ oxygen content of air at the Blower (kgO/m^3).

From equation (2-40) the airflow rate at the Blower can be calculated. With the airflow rate known, the total headloss in the system can be calculated and thus the pressure the blower must overcome. With the airflow rate and the pressure rating for the blower known, the correct blower can be chosen.

2.6.7.2.1 Influence of diffuser density and airflow rate on SOTE

As noted above, the airflow rate and diffuser density influence the SOTE. Usually the supplier will give a graph of the SOTE versus airflow rate / diffuser for different diffuser densities, typical to the one in Figure 5 below.

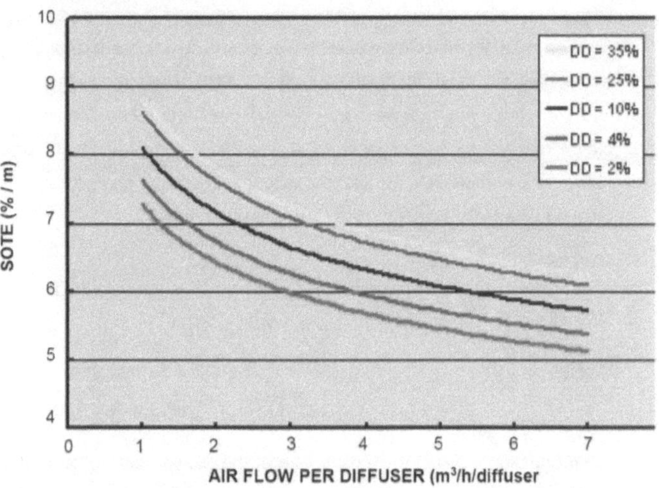

Figure 5: Oxygen Transfer efficiency under standard condition (SOTE) (%/m) versus airflow rate per diffuser (m³/hr /diffuser) (Wagner & Pöpel, 1998)

However, if this is not supplied, it needs to be included in the calculations in some fashion. Dold and Fairlamb (1999) have developed a method to do this mathematically and can be described as follows:

$$OTR_{std} = K_{La,std}C_{S,std}V \qquad (2\text{-}41)$$

Where:

$K_{La,std}$ $\quad=\quad$ mass transfer coefficient.

$$SOTE = {OTR_{std}}\big/{OSR_{std}} \times 100 \qquad (2\text{-}42)$$

This therefore adds another term to equation (2-34), which then becomes:

$$\frac{OTR_{site}}{OTR_{std}} = \left\{ \alpha\theta^{T-20}K_{LA,std}\frac{\left[\left(\frac{P_{site}-p_{site}}{P_{std}-p_{std}}\right)\left(\frac{P_{site}+73.53hf-P_{site}}{P_{site}-p_{site}}\right)\left(\frac{51.6}{31.6+T}\right)\beta C_{Sstd}F-C_L\right]}{C_s} \right\} \qquad (2\text{-}43)$$

Knowing the OTR$_{site}$ must meet the AOR i.e.

$$OTR_{site} = AOR$$

and OTE$_{site}$ is given by:

$$OTE_{site} = {OTR_{site}}\big/{OSR_{site}} = {OTR_{site}}\big/{\left(Con_{O2,site} \times Q_{Air,site}\right)}$$

$K_{La,std}$ must therefore be determined to meet the AOR and then the OTE_{site} can be determined. This means that the influence of the airflow rate and diffuser density on the K_{La} needs to be formulated.

Dold and Fairlamb note that the K_{La} is related to the superficial gas velocity given by:

$$K_{LaT} = C\, U_{SG}^{Y} \qquad\qquad\qquad (\,2\text{-}44\,)$$

Where:

C = constant dependant on the diffuser density;

U_{SG}^{Y} = superficial gas velocity (m /m /h)

= $Q_{Air,T}\big/A_{R};$

$Q_{Air,T}$ = air flow rate @ T °C;

A_{R} = area of aerated reactor (m^2).

In Equation (2-44), the constant C is dependent on the diffuser density via the following relationship:

$$C = k_1\, DD^{0.25} + k_2 \qquad\qquad\qquad (\,2\text{-}45\,)$$

Where:

DD = Diffuser Density (%)

= area diffusers/area reactor x 100;

k_1, k_2 = constants (/d).

In Equation(2-45), the constants k_1 and k_2 are specific for the diffuser type.

Thus, the 3 constants in the equations (2-44) and (2-45) above, Y, k_1 and k_2 need to be determined for the particular diffuser being used. This is done by curve fitting to the supplier's graph of SOTE versus airflow rate for different diffuser densities, typical to Figure 5. Once the constants are determined for the diffuser type, then the SOTE for any diffuser density and air flow rate can be calculated and so also the air flow rate to match the oxygen supply to the demand at any time over the day.

Note that once the aeration system has been designed, the diffuser density is fixed, and the value for C in equations (2-45) is also fixed. Hence, for a fixed diffuser density, only the constant C needs to be determined, not the individual k_1 and k_2.

From equation (2-44), SOTE is given as:

$$SOTE = C \left[\frac{Q_{Air,std}}{A_r}\right]^Y \left[\frac{C_{S,std} \, V}{Con_{O2,std} Q_{Air,std}}\right] {}^{100}\!/_{1000} \qquad (\,2\text{-}46\,)$$

Where:

1000	=	conversion g to kg ($C_{S,std}$ expressed as g/m^3 = mg/ℓ)
100	=	converts SOTE to %

In equation (2-46) Y and C are unknown, and must be determined. Y and C are determined through curve fitting equation (2-45) to the manufacturer's graph of SOTE vs Air Flow Rate for the set diffuser density. (Wagner & Pöpel, 1998)

Although the above method gives a mathematical procedure for calculating SOTE for any airflow, it is not commonly used in industry as the graph similar to Figure 5 is usually not available. In industry, manufacturers will supply a graph typical to Figure 6 below, unique to the diffuser type chosen. For this investigation the SSI ECD270 Disc Diffuser was chosen.

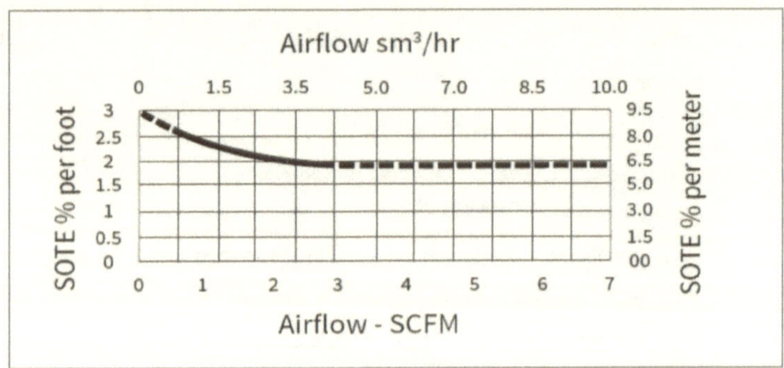

Figure 6: Airflow vs SOTE% (SSI ECD270 Disc Diffuser)

From the graph above, it is evident that the lower the airflow rate (sm^3/h) the higher the SOTE % per meter will be. For the diffuser chosen the practical air airflow rate vary between ± 0.75 and ±4.25 (sm^3/h) and SOTE % per meter vary between ±6.3 and ±8.2 (solid line). In industry a minimum SOTE % per meter will be chosen (for this investigation 7%) to which the manufacturer or designer must adhere. This will correspond to a maximum airflow rate per diffuser (from the graph ±2.5 sm^3/h) and depended on the total airflow rate required, the diffuser density can be calculated.

2.6.7.2.2 Calculating number of diffusers

The number of the diffusers is based on the diffuser density calculated in section 2.6.7.2.1. From this, the number of diffusers can then be calculated as follows:

$$No.of\ diffusers = \frac{Area_{Aer} \times DD}{Area_{diff}}$$

$$(\ 2\text{-}47\ )$$

Where:

$Area_{Aer}$ = Area of reactor aeration basin (m^2);

DD = Diffuser density (%);

$Area_{diff}$ = Area of diffuser (m^2).

2.6.7.3 Pressure Rating Calculation

As mentioned previously, in order for the correct blower to be chosen, the blower pressure rating must be known. To calculate the headloss (pressure the blower must overcome) through the system, the following must be considered:

- Headloss through the aeration pipes;
- Headloss through the diffusers;
- Submergence of diffusers;
- Fouling of diffusers.

2.6.7.3.1 Headloss through aeration pipes

Aeration piping consist of mains, valves, meters and other fittings that transport compressed air from blowers to the air diffusers. Because pressures are low, lightweight piping can normally be used. (Metcalf & Eddy, 2004)

The piping is sized so that the losses in the air headers and diffuser manifolds are low in comparison to losses in the diffuser.

Friction losses in air piping can be calculated using the Darcy-Weisbach equation:

$$h_L = \left[f \frac{L}{D} h_i \right]$$

$$(\ 2\text{-}48\ )$$

Where:

h_L =   friction loss (mm of water);

f =   dimensionless friction factor obtained from Moody diagram;

L =   equivalent length of pipe (m);

D =   pipe diameter (m);

h_i =   velocity head of air (mm of water).

(Metcalf & Eddy, 2004)

The friction factor for steel pipes carrying air can be approximated by the following equation (McGhee, 1991)

$$f = \frac{0.029(D)^{0.027}}{Q^{0.148}} \tag{2-49}$$

Where:

Q =   airflow (m^3/min) under prevailing pressure and temperature conditions;

D =   pipe diameter (m).

The headloss in straight pipes can be computed by substituting equation (2-49) into equation (2-48), giving:

$$h_L = 9.82 \times 10^{-8} \left(\frac{fLTQ^2}{PD} \right) \tag{2-50}$$

Where:

P =   air discharge pressure (atm);

T =   temperature in pipe (K) from following equation.

$$T = T_o \left({P}/{P_o} \right)^{0.283} \tag{2-51}$$

Where:

T_o =   ambient air temperature (K), maximum summer air temperature;

P_o =   ambient barometric pressure (atm).

(Metcalf & Eddy, 2004)

Losses in elbows, tees, valves ect., can be computed as a fraction of velocity head using headloss coefficient K values or as equivalent lengths of straight pipe with equation (2-52):

$$L = 55.4\ CD^{1.2} \qquad\qquad (2\text{-}52)$$

Where:

L = equivalent length of pipe (m);

D = pipe diameter (m);

C = resistance factor (table below).

Fitting	C factor
Long radius ell or run of standard tee	0.33
Medium radius ell or run of tee reduced by 25 %	0.42
Standard ell or run of tee reduced by 50 %	0.67
Tee through side outlet	1.33
Gate valves	0.25
Globe valves	2.00
Angle valves	0.90

Table 4: Resistance factors for fittings in aeration piping systems.

Meter losses can be estimated as a fraction of differential velocity head across the meter, depending on the type of meter. Losses in air filters, blower silencers and check valves must be obtained from equipment manufacturers but the following table can be used as a guide:

Device	Headloss (mm)
Air filters	13 - 76
Silencers:	
Centrifugal blower	13 – 38
Positive-displacement blower	152 – 216
Check valves	20 - 203

Table 5: Typical headlosses through air filters, blower silencers and check valves.

(Metcalf & Eddy, 2004)

From the above, it can be seen that calculations of gas (air) flow through pipelines are considerably more complicated than for liquid flow, due mainly to pressure-induced variations in the gas-stream density and velocity. For this reason, computer models have been developed for accurate calculations of gas flow through pipes.

If computer models are not available for the calculation, a rule of thumb method can be used. Through this method the pipe diameter is based on a constant pressure loss over a certain length of pipe. The Environmental Dynamic International (Technical Bulletin 145) suggest using a standard pressure drop of 3-inch wc / 100 ft (0.75 kPa / 30.5m) (Environmental Dynamics International, 2011).

The table below gives a relationship between the pipe diameter and velocities proposed to affect a constant pressure drop of approximately 0.75kPa per 30.5m of pipe length as well as the airflow rate:

Pipe Diameter (mm)	Velocity (m/s)	Airflow Rate (m^3/h)
50	8.0	68
90	10.5	204
110	12.8	425
150	17.3	1274
200	20.8	2633
250	24.4	4842
300	27.3	7646
350	30.0	10 390
400	30.0	13 572
450	30.0	17 177
500	30.0	21 205
550	30.0	25 659
600	30.0	30 536
700	30.0	41 563

Table 6: Velocity and airflow rate through aeration pipes.

As every WWTP is unique, the aeration pipework layout and associated pressures will be unique. In order to do the analysis for this investigation, the calculation was done on a 100m length of pipe (thus pressure drop of 2.46 kPa) and an additional pressure drop of 0.5 kPa to allow for all other losses (valves, bends ect.) through the pipe system. The total pressure drop through the pipe system was therefore taken as ±3.0 kPa. For calculation of the pipe size, Table 6 was used together with the airflow rates calculated.

2.6.7.3.2 Headloss through diffusers

For this investigation the SSI AFD270 (1mm) 9" Disc Diffuser supplied by SSI Aeration was chosen as the diffuser type. To calculate the headloss through the diffusers, the following graph (as supplied by the manufacturer) was used.

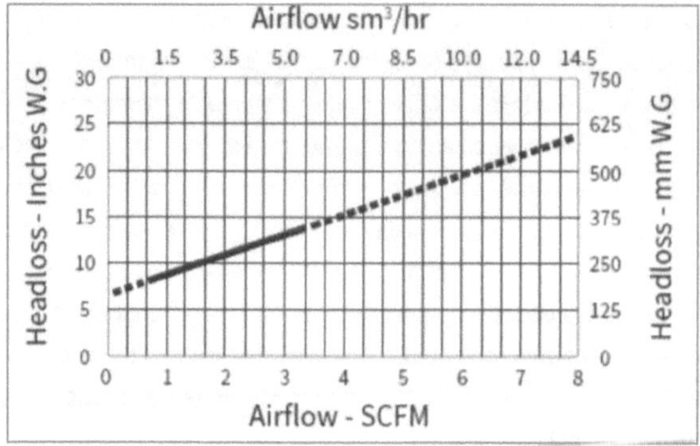

Figure 7: Headloss through diffuser (where W.G is water gauge pressure).

The graph represents the headloss in mmWG (millimeters water gauge i.e. 9.81 Pa) vs Airflow rate per diffuser (m³/hr). In order to ease the calculation process, the following equation was calculated (through graph fitting) to represent the graph:

$$Headloss = 33.19(Airflow\ rate) + 171.81 \qquad\qquad (2\text{-}53)$$

Equation (2-53) was therefore used to calculate the headloss through the diffusers.

2.6.7.3.3 Submergence of diffusers

As the diffusers are submerged beneath the water, the air pressure through the diffusers must overcome this water pressure in order to function. Dependent on the depth of submergence of the diffuser, the water pressure can be calculated with the following equation:

$$Headloss = 9.81 \times depth\ of\ submergence \qquad\qquad (\ 2\text{-}54\ )$$

2.6.7.3.4 Fouling of diffusers

Fouling of diffusers are a common occurrence in FBDA systems. Fouling is mainly due to solids backflow, salt buildup due to evaporation and biological growth on the diffusers. Although newer diffuser membrane technologies have reduced the amount of fouling on these systems, the design of the blower must allow for a certain amount of fouling to occur. If membrane pressure increases more than 25% above a clean membrane this is considered the upper limit of acceptable membrane performance (life). Routine maintenance or cleaning may reduce the membrane pressure and thus provide additional economic life. If no improvement in pressure after cleaning or maintenance, the economics of replacement should be reviewed. (Environmental Dynamics International, 2005). For this reason, the blower must allow for pressures 25% above the normal operating pressures of the diffusers. Looking at Figure 7 and an airflow rate of 2.5 sm^3/hr, the normal operating pressure of the diffusers is ± 250 mmWG (2.45 kPa). The additional pressure the fouling of the diffuser will therefore exert on the system is 2.45 x 25% = 0.62 kPa.

2.6.7.3.5 Total headloss through FBDA system

Taking the above into account, the blower's supplied pressure must therefore overcome this total headloss in order for the system to function.

As mentioned above, with the pressure and the airflow rate known the correct blower can be chosen for the system.

2.6.7.4 Blower Selection

With the airflow rate and pressure required known, the size (kW) of the blowers required can be calculated. The same as for the surface aerators, the conversion from airflow rate to size of blower will vary dependant on type of blower and manufacturer.

There are essentially three types of blowers commonly used for aeration i.e. centrifugal, rotary lobe positive displacement, and inlet guide vane-variable diffuser.

In WWTPs, the blowers must supply a wide range of airflows with a relatively narrow pressure range under varied site conditions. A blower usually can only meet one particular set of operating conditions efficiently. Due to the wide range of airflow and pressures that must be met at WWTPs, provisions have to be included in the blower system design to regulate or turn down the blowers. These methods include:

1) Flow blow-off or bypassing

2) Inlet throttling

3) Adjustable discharge diffusers

4) Variable speed drives

5) Parallel operations of multiple units.

(Metcalf & Eddy, 2004)

A relatively new blower design, inlet guide vane – variable diffuser, that was developed in Europe, mitigates some of the problems and considerations associated with standard centrifugal and positive-displacement aeration blowers. The design is based on a single-stage centrifugal operation that incorporates actuators to position the inlet guide vane and variable vane diffusers to vary blower airflow rate and optimise efficiency. The blowers are especially well suited to applications with medium to high fluctuations in inlet temperature, discharge pressure and flowrates. (Metcalf & Eddy, 2004)

For this reason, and because it is widely used in the South African industry, this single stage centrifugal type blowers with inlet guide vane – variable diffuser control was chosen for this investigation.

To calculate the power required for the blower, the performance curve for a centrifugal blower resembles the performance curve for a centrifugal pump and can be used. The performance curve typically is a falling-head curve where the pressure decrease as the inlet volume increases. Blowers are normally rated for standard conditions, defined as temperature of 20°C, a pressure of 760mm Hg and a relative humidity of 36%. Standard air has a specific weight of 1.20 kg/m^3. The density of air affects the performance of the centrifugal blower; any change in the inlet air temperature and barometric pressure will change the density of the compressed air. The greater the air density, the higher the pressure will rise. As a result, greater power is needed for the compression process. The blower selected must be of adequate capacity to deliver in all weather conditions. The blower power requirement for adiabatic compression is given by the following equation:

$$P_w = \frac{wRT_1}{29.7\,n\,e}\left[\left(\frac{p_2}{p_1}\right)^{0.283} - 1\right]$$

(2-55)

Where:

P_w	=	power requirement (kW);
w	=	weight of flow of air (kg/s);
R	=	universal gas constant for air
	=	8.314 J/(mol.K);
T_1	=	absolute temperature (K);
p_1	=	absolute pressure at inlet (atm);
p_2	=	absolute pressure at outlet (atm);
n	=	(k-1)/k = 0.283 for air;
29.7	=	constant for SI units' conversion;
550	=	ft.lb/s per hp;
e	=	efficiency (usual range for compressors is 0.70 to 0.90)
	=	0.80 for this investigation.

(Metcalf & Eddy, 2004)

To calculate the absolute pressure at the inlet of the blower (p_1), the altitude and temperature at the blower must be taken into account. The following equation represents this relationship:

$$p_1 = p_0 \exp\left(-\frac{Mg}{RT}h\right)$$

(2-56)

Where:

p_0	=	pressure at sea level (kPa);
m	=	molar mass of air
	=	0.2896 kg/mol;
g	=	gravitational acceleration;

R	=	universal gas constant for air
	=	8.314 J/(mol.K);
T_1	=	temperature at blower (K);
h	=	altitude (m).

To calculate the weight of flow of air (w) the following equation can be used:

$$w = Q_{Air,Blower} \times \rho \qquad (2\text{-}57)$$

Where:

| $Q_{Air,Blower}$ | = | Air flow rate required by blower (m³/s); |
| ρ | = | Air density (kg/m³). |

As the specific weight of air (w) in equation (2-55) is for standard conditions, the air density must be adjusted to allow for site conditions (at blower).

To calculate the density of air at other temperatures the following relationship can be used:

$$\rho_a = \frac{PM}{RT_a} \qquad (2\text{-}58)$$

Where:

P	=	atmospheric pressure
	=	1.01325 x 10⁵ N/m²;
M	=	mole of air
	=	28.97 kg/kg-mole;
R	=	universal gas constant;
T	=	temperature, K
	=	273.15 + °C.

(Metcalf & Eddy, 2004)

The following relationship can be used to compute the air density with a change in atmospheric pressure with elevation:

$$P_b = P_a exp\left[-\frac{gM(Z_b - Z_a)}{RT_b}\right] \qquad (2\text{-}59)$$

Where:

P_a	=	atmospheric pressure
	=	1.01325 x 10⁵ N/m²;
P_b	=	pressure to be calculated;

g =   gravitational acceleration;

M =   mole of air
        =   28.97 kg/kg-mole;

R =   universal gas constant;

Z =   elevation (m).

(Metcalf & Eddy, 2004)

Combing equations (2-58) and (2-59) gives:

$$\rho_b = \frac{P_a exp\left[-\frac{gM(Z_b - Z_a)}{RT_a}\right] M}{RT_b} \qquad (\,2\text{-}60\,)$$

With equation (2-60) the air density can be calculated at any temperature and elevation.

To compensate for humidity, the density of humid air can be calculated by treating it as a mixture of ideal gasses. The following equations can be used to calculate the density of humid air:

$$\rho_{humid\ air} = 1/v = \frac{\left(p/R_aT\right)(1+x)}{\left(1 + x\,{}^{R_w}/_{R_a}\right)} \qquad (\,2\text{-}61\,)$$

Where:

v =   specific volume of moist air per mass unit of dry air and water vapor (m^3/kg);

R_a =   286.9 – Individual gas constant for air (J/kg K);

R_w =   461.5 – Individual gas constant for water vapour (J/kg K);

x =   humidity ratio (kg water/kg air);

p =   pressure in the humid air (Pa).

The density of dry air can be expressed as:

$$\rho_{dry\ air} = p/R_aT \qquad (\,2\text{-}62\,)$$

Where:

$\rho_{dry\ air}$ =   density of dry air (m^3/kg)

Combining equations (2-61) and (2-62):

$$\rho_{humid\ air} = \frac{\rho_{dry\ air}(1+x)}{\left(1 + x\,{}^{R_w}/_{R_a}\right)} \qquad (\,2\text{-}63\,)$$

The gas constant ratio between water vapor and air is:

$$R_w/R_a = \frac{\left(461.5\,\frac{J}{kg}\,K\right)}{286.9\,\frac{J}{kg}\,K} = 1.609$$

Inserting this into equation (2-63) gives:

$$\rho_{humid\ air} = \frac{\rho_{dry\ air}(1+x)}{(1+1.609x)} \qquad\qquad (\,2\text{-}64\,)$$

(The Engineering Toolbox, 2019)

From equation (2-64) above, it can be seen that density of humid air will be less than dry air. The humidity ratio (x) is also relatively small (see table attached as Appendix B). For example, at a temperature of 35°C it is equal to 0.036756. The result is that the density of humid air is 97.9% as dense as dry air at 35°C, making it negligible to the calculation. The influence of humid air on air density was therefore ignored in this investigation.

With the air density converted from standard condition to site conditions, equation (2-55) can used to calculate the power requirement of the blower. With the total power requirement known, the correct size blowers can be chosen. As blower are only available in standard sizes (see section 2.8.1.8 below), the sizes of the blowers were chosen to best suit the power requirement calculated above.

With the correct aeration equipment chosen, it is important that the system is controlled with the correct control processes and devices in order to minimise the energy requirement.

2.7 Process and Aeration Control

Advanced process control and automation is becoming ever more important on WWTPs in order to better control plants and cut down wastage of energy. Various case studies have shown significant savings in operating cost when these systems have been employed. The ultimate purpose of any wastewater treatment plant is to satisfy the effluent requirement at the minimal possible cost. (Olsson, 2008)

As mentioned previously, the aeration system accounts for between 45 and 60% of the energy usage of an activated sludge plant. The control of the aeration system to ensure dissolved oxygen (DO) concentration remain at an optimum, therefore becomes very important when looking at energy optimisation. From section 2.6.4 it is evident that the cycle flow and load conditions will influence the biological oxygen requirement. The oxygen supply equipment must therefore adjust to these conditions in order to optimise energy usage. For this reason, aeration control systems were investigated during this study.

2.7.1 Aeration Control Systems

As dissolved oxygen is the most important variable in the activated sludge process from a biological treatment and energy consumption point of view, the set point and control of the DO concentration is crucial. It has been found that low DO concentrations produce MLSS with poor settling characteristics, known as bulking sludge, which can cause high concentrations of suspended solids in the final effluent and poor treatment performance. On the other hand, too high DO concentrations will constitute wastage of energy. Experience have shown that a DO concentration of $\pm$ 2 mg/ℓ is the optimum concentration. To satisfy the effluent requirement remains the main priority of any WWTP and the DO concentration set point must therefore be altered to satisfy this requirement but with the minimal energy input.

As DO concentrations can these days be measured very accurately with Dissolved Oxygen Probes, control systems can use the information to maintain the optimum DO concentrations in the aeration zones. As the DO probes play such an important role in the control of the aeration system, the accuracy of the information retrieved are very important. These probes must therefore be maintained and calibrated regularly. Another important property of the probe is the geographical position in the aeration zone itself, as DO levels in the aeration basin may vary significantly. Figure 8 below gives a typical presentation of the DO concentration variance that can be expected in a SSSA and FBDA system.

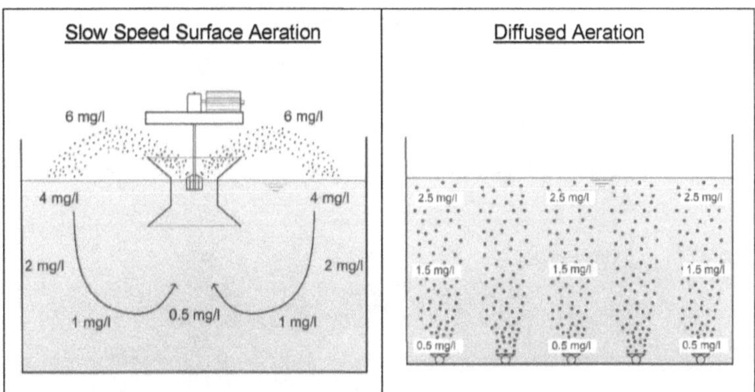

Figure 8: Typical DO Concentrations – SSSA vs FBDA

From the figure above it is evident that the DO concentration for SSSA systems vary significantly depending on the relative location to the aeration device. On the other hand, FBDA systems give a more uniform DO concentration throughout aeration basin. Controlling SSSA system via DO is therefore difficult and for this reason timer control and variable speed drive (VSD) systems are mostly implemented with SSSA systems. Although these control systems yield positive energy saving results, more operator input is required to keep the system functioning optimally. A complete automatic DO control system for SSSA systems is therefore difficult to implement.

Automatic DO control for FBDA systems have however proved to be very successful and various DO control systems have been developed in the past couple of years. Systems like the BioActive Response System (BARS) and the Bioprocess Intelligent Optimization System (BIOS) has been implemented with great success worldwide.

2.7.1.1 BioActive Response System (BARS)

2.7.1.1.1 Description

BARS is an aeration control system with the aim to keep DO concentrations in the aerobic basin at a certain set point by consuming the least possible energy. BARS aim to maintain the oxygen level needed for the biological processes, unaffected by increase of flow, temperature and depth of tanks. This happens automatically with BARS adjusting valves and the blower pressure set point to suit the required needs. With BARS it is possible to keep an almost linear DO level by measuring the airflow to each zone. This ensures that any pressure change in the system is not affecting the DO level in that zone. (Howden, 2017)

Figure 9 below gives a presentation of the BARS control philosophy.

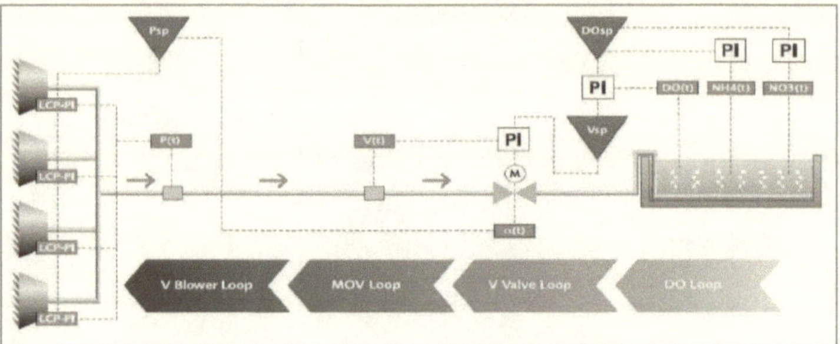

Figure 9: BARS Process Flow Diagram (Howden, 2017)

2.7.1.1.2 Control Philosophy

BARS is an automatic control system based on cascading DO Airflow Control Loop in each aeration basin. The basin modulating valves adjust in response to the oxygen demand, indicated by the DO probes, and cascade calculated flow set points. (Howden, 2017)

2.7.1.1.3 DO Control Loop

The DO Control Loop forms the heart of successfully controlling the process. The DO sensors continuously measure the dissolved oxygen level in each aeration zone. The DO controller output is then used to calculate the Airflow Set Point of the Airflow Control Loop. (Howden, 2017)

2.7.1.1.4 Airflow Control Loop

The Airflow Control Loop is the next step in the control process. By using a thermal mass flow transmitter, the flow controller ensures that the control valves are always in the correct position to maintain an optimal calculated Airflow Set Point through the air diffusers.

The flow meter continuously measures the amount of air through the air diffusers in the basin and adjust the modulating valve using a proportional-integral-derivative (PID) regulator. This ensures the air flow is dictated by the Flow Set Point (calculated from DO Loop) and therefore optimising and minimising the power consumption of the blowers. (Howden, 2017)

2.7.1.1.5 Most Open Valve (MOV) Control

BARS also monitor the valve positions in order to adjust the pressure set point to a level that ensures that all valves are kept as open as possible. This allows the blowers to produce only the airflow required by the process with minimal energy consumption in losses.

An integrated MOV Step/Wait Control regulator ensures the optimal and most efficient header pressure by continuously reading the "Average Valve Positions" of all the basins. Based on the Average Valve Position of the most open basin the MOV Step/Wait controller will adjust the Pressure Setpoint accordingly, optimising and minimising the power consumption. (Howden, 2017)

2.7.1.1.6 Pressure Control Loop

The Pressure Controller is fourth loop which controls the air flow discharged from the on-line blowers, keeping a constant main header pressure. By using the main header pressure transmitter, the controller ensures that the blower inlet guide vane (IGV) positions are always in the correct position to maintain an optimal header pressure set point, which is calculated by the MOV Control loop.

The Pressure Transmitter continuously measure the main header pressure and adjust the modulating IGV positions on the blowers using a PID regulator. This ensures the pressure load, dictated by the MOV Pressure Set Point, is equally shared on all the duty blowers. (Howden, 2017)

2.7.1.1.7 Ammonia Control

BARS also offer the option to control the dissolved oxygen level based on the load and speed of nitrification. The ammonium level is carefully monitored (by ammonia probes) and depending on the allowed value of ammonium level, the oxygen Set Point is adjusted to deliver the most cost-beneficial performance while remaining within the allowed limits. Typically, the oxygen level may stay within 0.5mg/ℓ to 2.5mg/ℓ when using this control which can result in major power savings. (Howden, 2017)

2.7.1.2 Bioprocess Intelligent Optimization System (BIOS)

2.7.1.2.1 Description

BIOS is a proprietary control algorithm, on-line process simulation program originally developed by Biochem Technology, Inc. to optimise the operation of the MLE process configuration.   Since the nitrification / de-nitrification sections of the MLE process configuration are an integral part of many other biological nutrient removal processes, the BIOS control system can be applied to other processes having the MLE component. (BioChem Technology Inc., 2010)

Figure 10 below gives a presentation of the BIOS control philosophy.

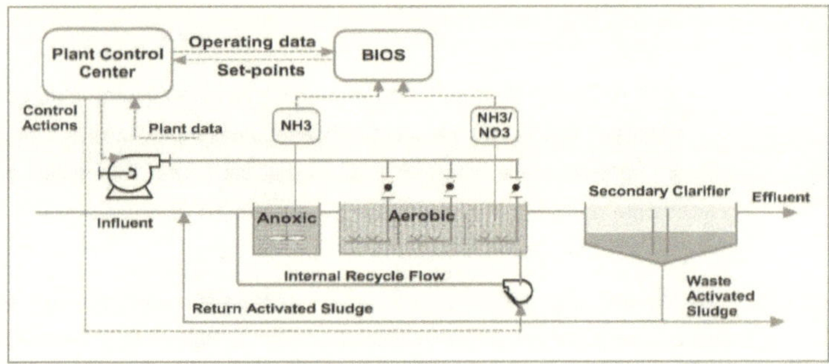

Figure 10: BIOS Process Flow Diagram (BioChem Technology Inc., 2010)

2.7.1.2.2 Control Philosophy

BIOS is a feed-forward optimisation system that conducts simulation calculations based upon on-line measurement of temperature, ammonia, nitrate and influent wastewater flow rate.   The system uses integrating process measurements with laboratory analytical results from the MLSS and use these as inputs to the algorithm. The BIOS simulation provides a continuous output of DO setpoints for the biological treatment process according to the load entering the bioreactor.   Additionally, the internal recirculation flow rate (IRQ) from the aerobic zone to the upstream anoxic zone in the MLE process or multi-zone biological nutrient removal (BNR) process is controlled to achieve optimal total nitrogen removal.   Using BIOS to control the biological nitrogen removal process produces low effluent total nitrogen concentration while minimising aeration energy consumption. (BioChem Technology Inc., 2010)

2.7.1.2.3 DO Control Theory

The BIOS DO Set-Point Controller calculates the optimal DO set-point(s) to treat the COD and ammonia loading. Optimising the DO set-point(s) decreases the aeration requirements and increases the oxygen transfer efficiency, ultimately resulting in energy savings.

An ammonia analyser located in the last anoxic zone provides the control system with the ammonia concentration entering the aerobic zones. Utilising the measured ammonia concentration along with other measured parameters such as flow rates, DO, and MLSS, the BIOS DO Set-Point Controller conducts biological and hydraulic simulations to predict how the DO concentration will affect the final effluent quality. The optimum DO set point(s) is the minimum DO concentration required to meet the effluent quality goals. An effluent ammonia analyser is used to confirm the simulation results and automatically adjust the model parameters if necessary. (BioChem Technology Inc., 2010)

2.7.1.2.4 IRQ Control Theory

The BIOS IRQ Set-Point Controller calculates the IRQ set-point required to maximise denitrification performance. Optimising the IRQ set-point results in several benefits including: 1) minimising effluent total nitrogen, 2) maximising oxygen credits generated in the anoxic zone(s) to reduce aeration requirements and 3) minimising IRQ pumping energy.

A nitrate analyser located at the IRQ pump inlet provides the control system with the nitrate concentration in the IR stream. The controller conducts iterated biological and hydraulic simulations that predict the effluent ammonia and nitrate concentrations under different IRQ rates. The optimum IRQ set-point is the value that results in the lowest effluent ammonia and nitrate concentration. (BioChem Technology Inc., 2010)

2.7.1.3 Aeration Control System Choice and Typical Energy Savings

Although both aeration control systems explained above has been proven to yield positive energy saving results, for this study the BARS was incorporated due to its simplicity and frequent use in the South African market. As energy savings with this system is difficult to predict, as it is dependent on the WWTP conditions, a percentage saving was incorporated. From case studies, one could conservatively expect a 15% saving on life cycle costs by installing DO basin control (Mynhardt, 2017) such as BARS. For this reason, an energy saving of 15% was assumed for this investigation.

This energy saving was incorporated into the operational cost of the cost modeling done for the FBDA system.

2.8 Cost Modelling

With the design and sizing of the infrastructure and equipment associated with the different aeration technologies (SSSA vs FBDA) completed, it was possible to complete the costing for each system. For the cost modelling, the capital, operational and maintenance costs were considered.

2.8.1 Capital Cost

Capital costs are fixed, one-time expenses incurred with the purchase and/or construction of equipment and infrastructure. Due to the vast range of equipment and infrastructure associated with the construction of a WWTP, estimating capital cost for plants can be a very time-consuming exercise. For this reason, cost functions (mathematical formulas used to represent how expenses will change with different input parameters) were used to calculate the capital cost.

By implementing cost functions for different types of infrastructure and equipment, the capital cost of any size infrastructure or equipment (within constraints) can be estimated. To calculate the cost functions, historical costing data on implemented projects (2010 to 2018) with different size infrastructure was gathered and escalated to 2019. This data was entered into Microsoft Excel and graphs were plotted portraying capital cost vs size of infrastructure or equipment. By applying the least-squares regression method to this data, the cost functions for the different infrastructure and equipment were calculated as detailed below.

2.8.1.1 Biological Reactors (SSSA)

The capital cost of biological reactors for slow speed surface aeration were based on the following:

➢ Rectangular, steel reinforced concrete structure with a water depth of 4.50m and freeboard of 800mm (total wall height 5.30m). Structure semi-submerged below ground level.

➢ Structure includes steel reinforced concrete walkways and platforms to support mixers and surface aerators.

➢ Sizing constraints enforced on reactors due to practicality of construction - volume between 1 000 and 16 000m^3.

From the graph below the following cost function for the Biological Reactor (SSSA) can be obtained.

Cost Function: $Cost = 30.08V^{0.656} \times 1000$

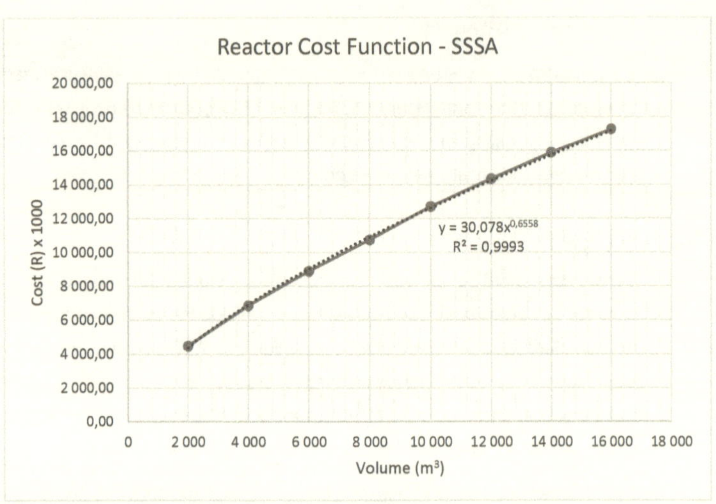

Figure 11: Reactor Cost Function – SSSA graph

2.8.1.2 Biological Reactors (FBDA)

The capital cost of biological reactors where fine bubble diffused aeration will be implemented were based on the following:

> Rectangular, steel reinforced concrete structure with a water depth of 4.50m and freeboard of 500mm (total wall height 5.00m). Structure semi-submerged below ground level.

> Structure includes steel reinforced concrete walkways and platforms to support mixers (no aerator platforms required).

> Sizing constraints enforced on reactors due to practicality of construction - volume between 1 000 and 16 000m³.

From the graph below the following cost function for the Biological Reactor (FBDA) can be obtained.

Cost Function: $Cost = 26.35V^{0.663} \times 1000$

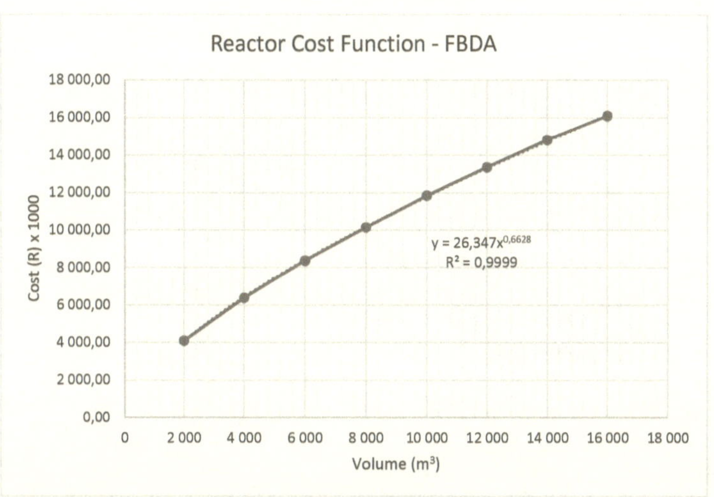

Figure 12: Reactor Cost Function – FBDA graph

2.8.1.3 Secondary Settling Tanks

The secondary settling tank cost is based on the following:

- ➢ Circular, steel reinforced concrete structure with a minimum water depth of 4.00m and freeboard of 800mm. Structure semi-submerged below ground level.
- ➢ Sloped floor bottom type with scraper mechanism and sludge hopper.
- ➢ Sizing constraints enforced on tanks due to practicality of construction - diameter between 15m and 35m.

From the graph below the following cost function for the Secondary Settling Tank can be obtained.

Cost Function: $Cost = 137.67\emptyset^{0.957} \times 1000$

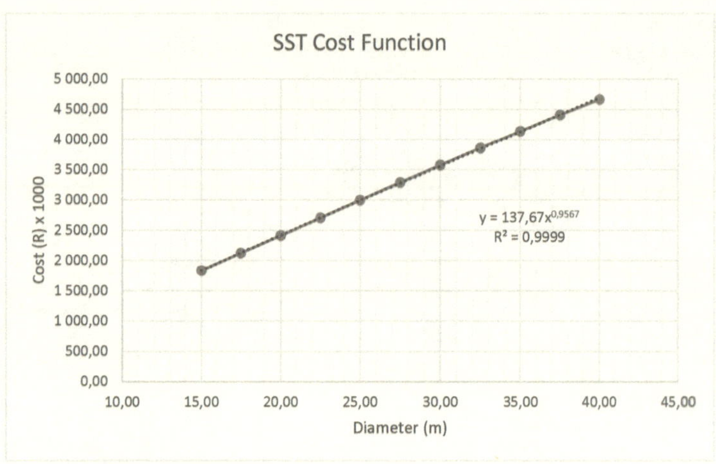

Figure 13: Secondary Settling Tank Cost Function graph

2.8.1.4 Blower House Building

As described in section 2.5.6, the costing of the blower house building is based on the size and type of the blower equipment installed. The cost of the building was based on the following:

➢ Steel reinforced concrete frame building with masonry infill walls.

➢ Roof covering – corrugated iron roof sheets

➢ Area of building dependent on number and size of blowers.

Sizing constraints enforced on building due to practicality and workspace – minimum area of 75m^2.

From the graph below the following cost function for the Blower House Building can be obtained.

Cost Function: $Cost = 85.12(Area)^{0.514} \times 1000$

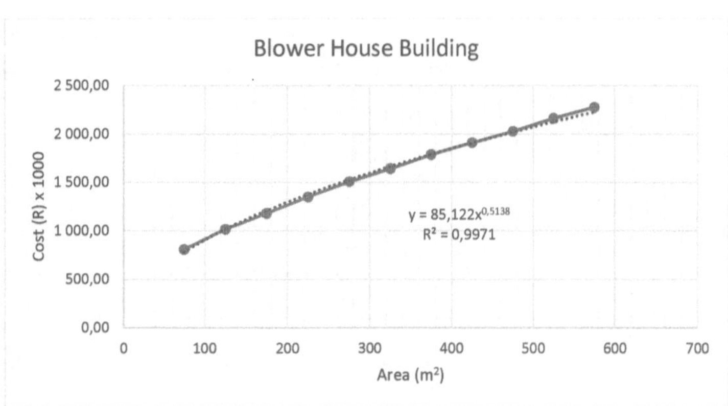

Figure 14: Blower House Building Cost Function graph

2.8.1.5 Slow Speed Surface Aeration – Surface Aerators

The slow speed surface aeration equipment is based on the size of the surface aerator equipment required and typically used in the South African market. Slow Speed Surface Aerators with a maximum rotational speed of 45 rpm was chosen. Sizing constraints for the aerators are between 10 kW and 110kW.

From the graph below the following cost function for the Slow Speed Surface Aeration can be obtained.

Cost Function: $Cost = 31.89(kW)^{0.687} \times 1000$

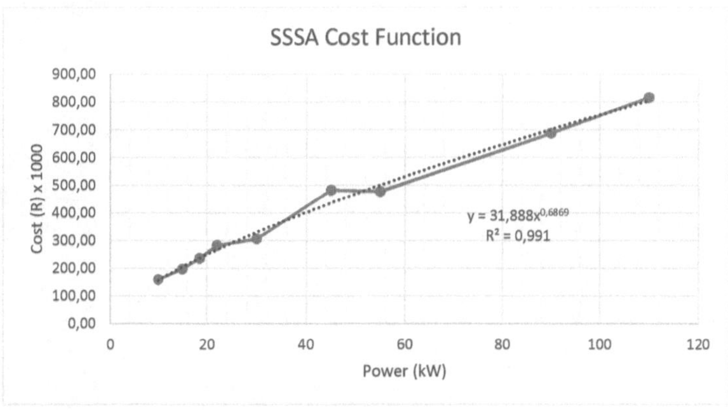

Figure 15: Slow Speed Surface Aeration Cost Function graph

As mentioned in section 2.6.6, aerators are only available in standard sizes. Table 7
below gives a list of the standard aerator sizes available.

Standard Aerator Sizes	
10 kW	55 kW
15 kW	65 kW
22 kW	80 kW
30 kW	95 kW
45 kW	110 kW

Table 7: Standard aerator sizes

With reference to the table above, the capital cost of the surface aeration equipment
was based on the aerator sizes chosen and is slightly more than the total power
requirement calculated from equation (2-33). The power requirement calculated
from equation (2-33) was however used for the calculation of the operational cost,
as it is assumed that the aeration control system will regulate the system and only the
energy required will be consumed.

2.8.1.6 Fine Bubble Diffused Aeration – Blower System

The costing of the blower system is based on the GTB type Turbo Compressor as supplied by Next Turbo Technologies. Figure 16 below is a photo of a typical GTB Turbo Compressor.

Figure 16: Typical detail GTB Turbo Compressor.

The costing of the blower system includes the compressors (blowers), electric-drive-motor, inlet silencer, discharge diffuser, blow-off valve and local control panel.

From the graph below the following cost function for the Fine Bubble Diffused Aeration – Blower System can be obtained.

Cost Function: $Cost = 886.53(kW)^{0.345} \times 1000$

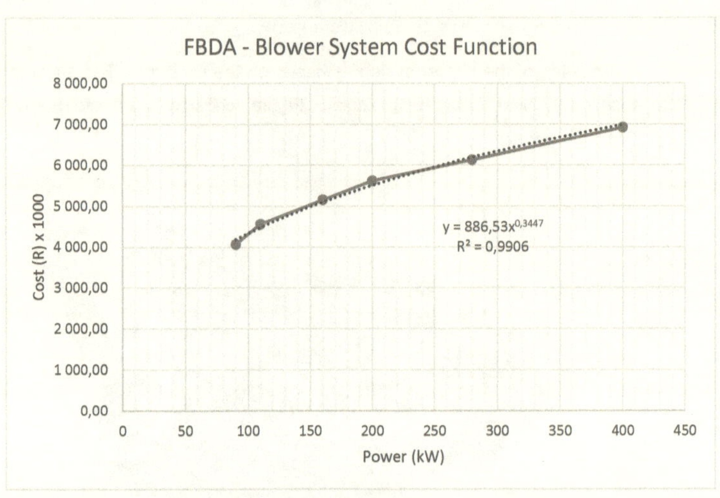

Figure 17: Fine Bubble Diffused Aeration – Blower System Cost Function graph

As mentioned in section 2.6.7.4 blowers are only available in standard sizes. Table 8 below gives a list of the standard blower sizes available.

Standard Blower Sizes	
Type	Power Output
GT-B-T10	50 kW, 75 kW, 90 kW
GT-B-T20	110 kW, 160 kW
GT-B-T30	200 kW, 250 kW, 315 kW
GT-B-T40	400kW, 450 kW

Table 8: Standard Blower Sizes

As a result, the capital cost of the blower system equipment was based on the blower size chosen. As the blower system must also make allowance for equipment failure, a duty/standby blower configuration is required. For this investigation it was decided that for systems with a power requirement of less than 160 kW, a one (1) duty – one (1) standby configuration will be installed. For systems with power requirements of more than 160 kW, a two (2) duty – one (1) standby configuration will be installed.

The capital cost for the system was therefore based on the equipment installed and is more than the total power requirement calculated from equation (2-55). The power requirement calculated from equation (2-55) was however used for the calculation of the operational cost, as it is assumed that the aeration control system will regulate the system and only the energy required will be consumed.

2.8.1.7 Main Aeration Header

As mentioned previously, the air pressure required for aeration by diffusers are quite low and therefore lightweight materials can be used. However, as it is good practice for aeration pipework to be installed above ground, for easy leak detection, stainless steel piping is normally the material of choice. As mentioned in section 2.6.7.3.1, every WWTP is unique and therefore the aeration header length will be different to each site. Therefore, in order to do a cost calculation, it was decided to allow for a 100m aeration header length for each plant. The cost function below is therefore based on this assumption with ranging pipe diameters.

$$Cost = 2322.5(Dia)^{0.802}$$

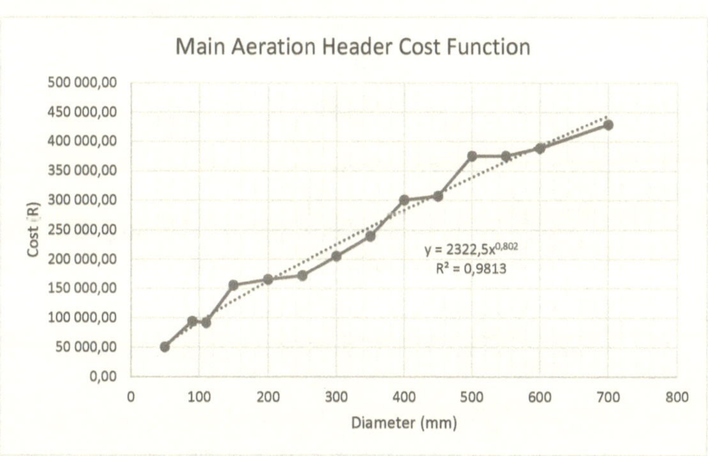

Figure 18: Main Aeration Header Cost Function graph

2.8.1.8 Fine Bubble Diffused Aeration – Diffuser System

The costing of the diffuser system is based on the SSI AFD270 (1mm) 9'' Disc Diffuser. Figure 19 below is a datasheet of this diffuser as supplied by the manufacturer.

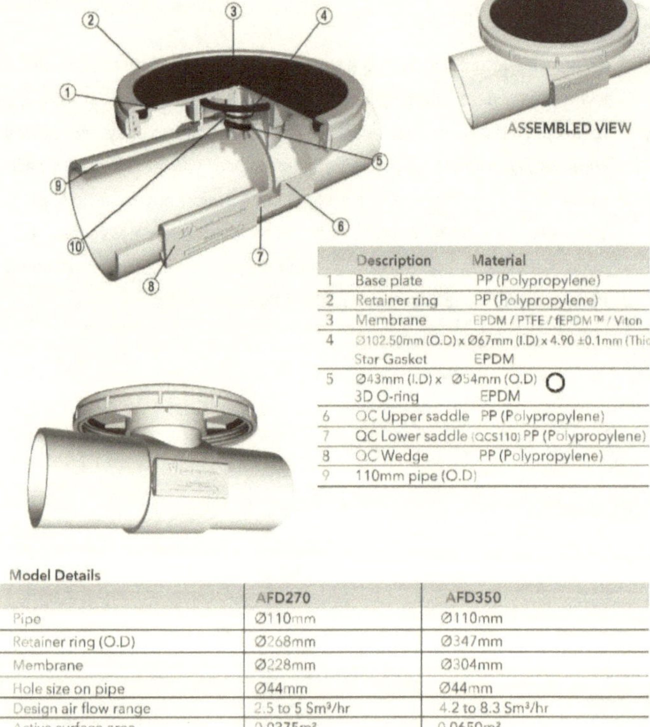

	Description	Material
1	Base plate	PP (Polypropylene)
2	Retainer ring	PP (Polypropylene)
3	Membrane	EPDM / PTFE / fEPDM™ / Viton
4	Ø102.50mm (O.D) x Ø67mm (I.D) x 4.90 ±0.1mm (Thick) Star Gasket	EPDM
5	Ø43mm (I.D) x Ø54mm (O.D) 3D O-ring	EPDM
6	QC Upper saddle	PP (Polypropylene)
7	QC Lower saddle (QCS110)	PP (Polypropylene)
8	QC Wedge	PP (Polypropylene)
9	110mm pipe (O.D)	

Model Details

	AFD270	AFD350
Pipe	Ø110mm	Ø110mm
Retainer ring (O.D)	Ø268mm	Ø347mm
Membrane	Ø228mm	Ø304mm
Hole size on pipe	Ø44mm	Ø44mm
Design air flow range	2.5 to 5 Sm³/hr	4.2 to 8.3 Sm³/hr
Active surface area	0.0375m²	0.0650m²

Figure 19: SSI AFD270 (1mm) 9'' Disc Diffuser datasheet.

The costing of the diffuser system includes all diffusers, uPVC piping inside the aeration basin, diffuser accessories to connect diffusers to pipes, stainless steel support stands and moisture purge system.

From the graph below the following cost function for the Fine Bubble Diffused Aeration – Diffuser System can be obtained.

Cost Function: $Cost = 1173.8(SOR) + 260769$

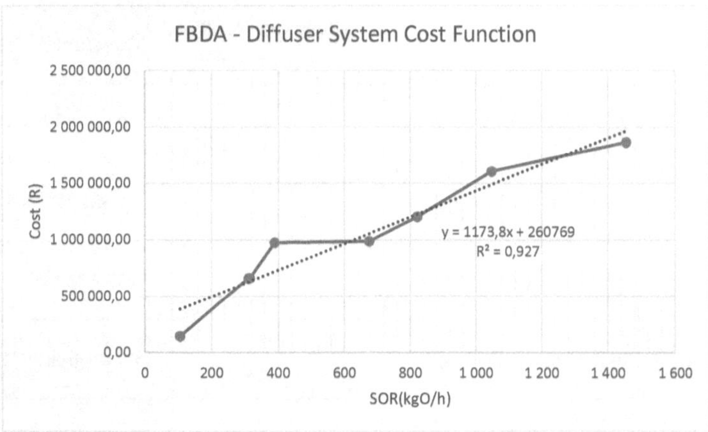

Figure 20: Fine Bubble Diffused Aeration – Diffuser System Cost Function graph

2.8.1.9 Electrical Equipment

The above-mentioned cost functions do not include the electrical and control equipment infrastructure. As the electrical and control equipment is unique to each plant layout, it was decided to calculate the electrical equipment cost based on a percentage of the cost of the other infrastructure. Based on similar completed projects the percentage cost of the electrical equipment is as follows:

> ➤ Slow Speed Surface Aeration 15%
> ➤ Fine Bubble Diffused Aeration 7.5%

2.8.1.10 Cost Functions

Table 9 below shows a summary of the cost functions for the infrastructure and equipment described above.

Unit Operation	Parameter (units) Range of validity	Cost function (R)
Biological Reactor (SSSA)	Volume per m³ 1 000 – 16 000m³	$30.08V^{0.656} \times 1000$
Biological Reactor (FBDA)	Volume per m³ 1 000 – 16 000m³	$26.35V^{0.663} \times 1000$
Secondary Settling Tanks	Diameter per m 15 – 35m	$137.67\emptyset^{0.957} \times 1000$
Blower House Building	Area per m² Min = 100m²	$85.12(Area)^{0.514} \times 1000$
Slow Speed Surface Aeration – Surface Aerators	Power per kW 10 – 110kW	$31.89(kW)^{0.687} \times 1000$
Blower System	Power per kW	$886.53(kW)^{0.345} \times 1000$
Main Aeration Header	Pipe per 100m	$2322.5(Dia)^{0.802}$
Diffuser System	SOR in kgO/hr	$1173.8(SOR) + 260769$

Table 9: Cost functions

With the cost functions known, it was possible to calculate the capital cost associated with the different infrastructure components and equipment.

2.8.2 Operational Cost

When looking at the operation cost of a WWTP, the electrical cost is by far the biggest contributor. The energy consumption between the different aeration technologies (SSSA vs FBDA) therefore played a major role in the outcome of the cost viability between the two technologies.

For the calculation of the electrical cost, the "2019/20 Schedule of Standard Prices for Eskom Tariffs" as produced by Eskom was used. The "Megaflex – Local Authority (Transmission zone ≤ 300km) costing template was used, see Appendix C.

In order to incorporate the "Peak", "Standard" and "Off peak" tariffs the following graph as supplied by Eskom was used:

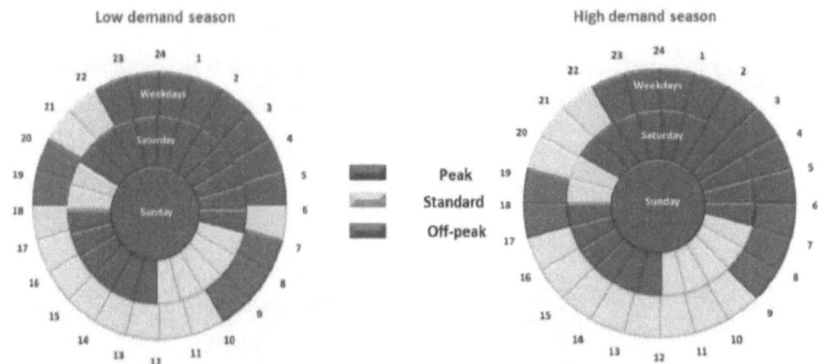

Figure 21: Low and high demand season time-of-use periods (Eskom, 2019)

With the "Megaflex" data and using the information in Figure 21 above, it was possible to calculate the total electrical cost. The electrical cost is essentially broken down into four (4) categories i.e. Active energy charge, Transmission network charge, Service charge and Administration charge. The Active energy charge is subject to "high and low demand season" as well as "Peak, Standard and Off-peak" times. Therefore, to calculate the Active energy charge, a yearly average of the hours and charges were calculated for the high and low demand season. The following table gives a summary of the charges used:

Description	Charge
Energy charge / kW / year (high demand)	R2 353.50
Energy charge / kW / year (low demand)	R4 407.34
Transmission network charge	R 11.20 / kVA / month
Service charge	R220.24 / day
Admin charge	R 99.28 / day

Table 10: Electricity Charges

Based on this information the total electrical cost for both aeration technologies were calculated.

2.8.3 Maintenance Cost

For the calculation of the total maintenance cost, the typical maintenance cost of civil and mechanical infrastructure as well as the lifespan for each piece of equipment was evaluated. Based on previous experience on similar completed projects, the following table gives the maintenance cost and lifespan of the equipment and this was used for this investigation.

Description	Maintenance Cost (% of capital cost)	Lifespan before replacement
Biological Reactors (Civil)	1%	50 years
Biological Reactors (Mechanical and Electrical)	4%	20 years
Secondary Settling Tanks (Civil)	1%	50 years
Secondary Settling Tanks (Mechanical and Electrical)	4%	20 years
Blower House	1%	50 years
Surface Aerators	4%	20 years
Blower System	4%	20 years
Diffuser System	4%	7 years

Table 11: Estimated maintenance cost and lifespan of equipment.

2.8.4 Comparison of Total Cost

From the information presented in Table 11 above, it was decided that a 20-year life cycle cost comparison between the different aeration technologies will be done. The following will be taken into account:

- Interest and redemption on total capital cost. For this investigation it was assumed the money required will be loaned at the prime lending rate (10.25% p.a.) and must be paid back over 20 years.
- Maintenance cost (civil, mechanical and electrical) over 20 years with an escalation rate of 7.00% p.a.
- Electrical operating cost over 20 years with an escalation rate of 10.00% p.a.

➤ Replacement cost of diffuser system. As it is assumed in this investigation that the diffuser system has a lifespan of 7 years, the cost of replacing the diffuser system two times over the 20-year period (including 7.00% escalation p.a. on the cost).

In order to eliminate the time value of money, the Net Present Value (NPV) method was implemented based on the following equation and parameters:

$$NPV = \sum_{t=1}^{T} \frac{C_t}{(1+r)^t} - C_0 \qquad\qquad (\text{ 2-65 })$$

Where,

C_t : Net cash inflow during the period

C_0 : Total initial investments

r : Discount rate

t : Number of time periods

(Investopedia, 2019)

Prime Lending Rate (June 2019)	:	10.25% p.a.
Escalation rate	:	7.00% p.a.
Power Operating Cost Escalation	:	10.00% p.a.
Number of time periods	:	20 years

The NPV was used to compare the different aeration technologies and the results and analysis of this can be found in section 4: RESULTS AND DISCUSSION.

2.9 Conclusion

With the ever-increasing cost of electricity in South Africa and the power utility, Eskom, at the brink of collapsing, it is of utmost importance that everything possible should be done to lower electrical consumption on WWTPs. As aeration is by far the largest consumer of electricity on WWTPs, choosing the correct aeration technology will not only decrease electrical cost but also ease the load on a struggling electrical grid.

3. METHODOLOGY

3.1 Influent and Site Data

The inflow volume (normally measured in Mℓ/d) to the WWTP as well as the characteristics of this influent (as described in section 2.3) is the starting point for any WWTP design. The more accurately these inputs can be measured or estimated, the more accurate the design of the WWTP will be. For upgrading of existing plants, it is recommended that sampling campaigns be conducted to measure these inputs and the necessary adjustment be made for the upgraded inflow. For new plants, data from surrounding plants can be gathered and / or calculated assumptions can be made to establish the type and inflow of wastewater to the new plant.

For this investigation, the upgrading of "Kelvin Jones Wastewater Treatment Works" was chosen as a sample project. The Kelvin Jones Wastewater Treatment Works is situated near Uitenhage in the Nelson Mandela Bay Metro, Eastern Cape Province, South Africa. The plant currently has a capacity of 24 Mℓ/d and is operated in a UCT process configuration. The plant is divided in two trains, a 10 Mℓ/d and 14 Mℓ/d respectively. Currently the flow to the plant is ± 16 Mℓ/d.

The design process for the upgrading of the plant is currently underway and it is proposed that it be done in two phases. Phase 1 of the upgrade will increase the plant's capacity to 30 Mℓ/d (17.5 Mℓ/d for train no. 1 and 12.5 Mℓ/d for train no. 2), treating raw wastewater. Phase 2 of the upgrade will increase the plant's capacity to 50 Mℓ/d (30 Mℓ/d for train no. 1 and 20 Mℓ/d for train no. 2), treating settled wastewater. Upgrading of the aeration system is one of the main systems that will form part of the upgrade. The current aeration system makes use of slow speed surface aeration but it is proposed to change this to a fine bubble diffused aeration system. The viable of this change is a critical element of the design and must be investigated.

As part of the design, a composite sampling campaign was conducted over a seven-day period from the 12th - 19th of February 2017. The results are summarised in Table 12 below. Due to a limited budget and time constraints, not all constituents could be measured during the sampling campaign and certain assumption had to made to complete the dataset. The legend underneath Table 12 gives an indication of the data used and assumptions made.

INFLUENT			
PARAMETER	UNIT	RAW WW	SETTLED WW
Current Flow	Mℓ/d	16.000	15.872
Future Flow (Phase 1)	Mℓ/d	30.000	29.760
Estimated Population (Future Flow)	persons	250 000	248 000
COD (S_{ti})	mgCOD/ℓ	672.000	398.000
COD Filtrate	mgCOD/ℓ	178.080	178.080
VFA	mg/ℓ	26.208	25.790
TOC	mgC/ℓ	224.000	133.000
TKN (N_{ti})	mgN/ℓ	53.760	46.105
FSA	mgN/ℓ	40.430	50.470
TP (P_{ti})	mgP/ℓ	12.800	10.880
OP	mgP/ℓ	8.100	8.100
TSS	mgTSS/ℓ	343.000	89.180
ISS	mgISS/ℓ	47.700	9.500
Influent pH		7.330	7.330
Influent alkalinity	mg/ℓ as $CaCO_3$	387.000	387.000
EFFLUENT			
COD	mgCOD/ℓ	47.040	47.040
VFA	mg/ℓ	0.000	0.000
TOC	mgC/ℓ	15.680	15.680
TKN	mgN/ℓ	1.838	2.135
FSA	mgN/ℓ	0.423	0.376
TP	mgP/ℓ	0.218	0.218
OP	mgP/ℓ	0.218	0.218
TSS	mgTSS/ℓ	0.000	0.000
ISS	mgISS/ℓ	0.000	0.000
PST underflow	%	0.800	0.800
FRACTIONS			
$f_{S'us}$	mgCOD / mgCOD	0.070	0.117
$f_{S'up}$	mgCOD / mgCOD	0.150	0.040
$f_{S'bs}$	mgCOD / mgCOD	0.195	0.324
$f_{Sb's}$	mgCOD / mgCOD	0.250	0.385
$f_{Sbs'a}$	mgCOD / mgCOD	0.200	0.200
$f_{Sbs'f}$	mgCOD / mgCOD	0.800	0.800
f_{cv}	mgCOD / mgVSS	1.481	1.481
$F_{SR'S}$	%	40	

$f_{N'a}$	mgN / mgN	0.750	0.880
$f_{N'ous}$	mgN / mgN	0.030	0.041
$f_{Sus'N}$	mgN / mgCOD	0.034	0.034
$f_{Sup'N}$	mgN / mgCOD	0.068	0.068
$f_{Sb'N}$	mgN / mgCOD	0.009	0.008
$F_{SR'N}$	%	15	
$f_{P's}$	mgP / mgP	0.750	0.880
$f_{P'ous}$	mgP / mgP	0.017	0.020
$f_{Sus'P}$	mgP / mgCOD	0.005	0.005
$f_{Sup'P}$	mgP / mgCOD	0.020	0.020
$F_{SR'P}$	%	15	
UPO			
f_{cv}	mgCOD / mgVSS	1.481	1.481
fn	mgN/mgVSS	0.100	0.100
fc	mgC/mgVSS	0.518	0.518
fp	mgP/mgVSS	0.025	0.025
USO			
f_{cv}	mgCOD / mgVSS	1.493	1.493
fp	mgP/mgVSS	0.000	0.000
FBSO			
f_{cv}	mgCOD / mgVSS	1.420	1.420
SOLIDS			
TSS	mgTSS/ℓ	343	89.180
VSS	mgVSS/ℓ		
ISS	mgISS/ℓ	47.7	9.5
OTHERS			
μ_{A20}	/d	0.45	0.45
Alk	mg/ℓ as $CaCO_3$	200	200
Nitrification safety factor		1.250	1.250
Maximum polyphosphate P Content	mgP/mgPAOVSS	0.355	0.355
Anaerobic reactor mass fraction		0.120	0.120
Number of anaerobic compartments	No.	3	3
DSVI	mℓ/g	120.0	120.0

PDWF Factor relative to ADWF		1.707	1.709
PWWF Factor relative to PDWF		1.500	1.500
Peak OUR factor		1.250	1.250
Dissolved Oxygen in Reactor	mg/ℓ	2.000	2.000
SITE CONDITIONS			
T_{min}	°C	14	
T_{max}	°C	22	
Altitude	m	40.0	
α (SSSA)		0.80	
β (SSSA)		0.95	
θ (SSSA)		1.024	
R (oxygen transfer rate, SSSA)	kgO/kWh	2.00	
α (FBDA)		0.50	
β (FBDA)		0.95	
θ (FBDA)		1.024	
Minimum OTE	%/m	7.00	
Diffuser height	m	0.25	
Diffuser diameter	m	0.23	
Maximum airflow per diffuser (from graph)	m^3/h	2.5	
Max temperature at Blower	°C	36	
Reactor depth (water level)	m	4.50	

Table 12: Constitutes of influent as determined by sampling campaign.

Legend			
Measured		----	------
Typical values from literature (Table 3.2 of Characterization of Municipal Wastewater, (Wentzel & Ekama, 2003))		------	------
Typical values from literature and assumptions		------	------

As a sample viability calculation, the Kelvin Jones Raw wastewater inflow characteristics were used together with an inflow of 30 Mℓ/d to mimic phase 1 of the upgrade proposed at the Kelvin Jones WWTP. In order to see the difference between the different process configurations, both UCT and MLE process configurations were calculated. The calculation example is detailed below.

3.2 Kelvin Jones Process Infrastructure Design

With the Kelvin Jones input data portrayed in section 3.1 above, each infrastructure component of the WWTP was sized.

3.2.1 Balanced MLE Biological Reactor Design

Calculating volume of biological reactor from equation (2-4):

$$V_p = \frac{Q_{i,ADWF}\,S_{ti,Reactor}}{X_t}\,[A]$$

Parameters	Value (MLE)
$Q_{i,ADWF}$	$30.00\ M\ell/d$
$S_{ti,Reactor}$	$672\ mg/\ell$
X_t	$5.20\ kgTSS/m^3$
$[A]$	4.12
R_s	$12.15\ days$

Results	Value (MLE)
$V_p\ (Total)$	$15\ 961\ m^3$
Number of units	1
$V_p\ (Unit)$	$15\ 961\ m^2$

Calculating total oxygen demand from equation (2-5):

$$FO_t = FO_c + FO_n - FO_d$$

Parameters	Value (MLE)
FO_c	$11\ 203\ kgO_2/day$
FO_n	$4\ 605\ kgO_2/day$
FO_d	$2\ 522\ kgO_2/day$

Results	Value (MLE)
FO_t	$13\ 286\ kgO_2/day$

and equation (2-6):

$$OUR = {FO_t}/{V_{Aer}} \times 24$$

Parameters	Value (MLE)
FO_t	13 286 kgO_2/day
V_{Aer} (total)	10 318 m^3

Results	Value (MLE)
OUR	53.65 $mgO_2/(\ell.h)$

Calculating balanced sludge age from equation (2-7):

$$R_{sBalMLE} = \frac{C.E_{MLE} + D - A.B + A.S_f K_{2T}\frac{Y_H}{\mu_{AmT}} - E_{MLE}.f_n\left[A.Y_H + S_{ti}\frac{f_{Srup}}{f_{cvUPO}}\right]}{A(B.b_{HT} + K_{2T}Y_H) - A.S_f b_{AT} K_{2T}\frac{Y_H}{\mu_{AmT}} - b_{HT}(C.E_{MLE} + D) + E_{MLE}.f_n b_{HT}\left[A.Y_H f_H + S_{ti}\frac{f_{Srup}}{f_{cvUPO}}\right]}$$

Parameters	Value (MLE)
A	524.160
B	0.029
C	50.576
D	4.545
E_{MLE}	0.875

Results	Value (MLE)
$R_{sBalMLE}$	12.15 $days$

3.2.2 Balanced UCT Biological Reactor Design

Calculating volume of biological reactor from equation (2-8):

$$V_p = \frac{Q_{i,ADWF} S_{ti,Reactor}}{X_t} [A]$$

Parameters	Value (UCT)
$Q_{i,ADWF}$	$30.00 \ M\ell/d$
$S_{ti,Reactor}$	$672 \ mg/\ell$
X_t	$4.30 \ kgTSS/m^3$
$[A]$	5.99
R_s	$13.43 \ days$

Results	Value (UCT)
$V_p \ (Total)$	$27\,872 \ m^3$
Number of units	2
$V_p \ (Unit)$	$13\,936 \ m^2$

Calculating total oxygen demand from equation (2-5):

$$FO_t = FO_c + FO_n - FO_d$$

Parameters	Value (UCT)
FO_c	$10\,611 \ kgO_2/day$
FO_n	$4\,423 \ kgO_2/day$
FO_d	$2\,422 \ kgO_2/day$

Results	Value (UCT)
FO_t	$12\,612 \ kgO_2/day$

and equation (2-6):

$$OUR = {FO_t}/{V_{Aer}} \times 24$$

Parameters	Value (UCT)
FO_t	$12\,612\ kgO_2/day$
$V_{Aer}\ (total)$	$14\,999\ m^3$

Results	Value (UCT)
OUR	$35.04\ mgO_2/(\ell.h)$

Calculating balanced sludge age from equation (2-9) and (2-10):

$$LHS = D_{p1} = A\left[B\left(1-\frac{\%}{100}\right) + (1-F)K'_{2T}\left\{1 - \frac{S_f\left(b_{AT}+\frac{1}{R_S}\right)}{\mu_{AMT}} - f_{xa}\right\}\frac{Y_H R_S}{1 + b_{HT}R_S}\right]$$

Parameters	Value (UCT)
A	524.160
B	0.029
F	0.207
R_s	13.43 days

Results	Value (UCT)
LHS	32.773

$$RHS = D_{p1} = \frac{a_{prac}O_a + sO_s}{2.86} + \frac{\left(a_{prac}+s\right)}{\left(a_{prac}+s+1\right)}(C - N_s) = D_{UCT} + E_{UCT}G_{UCT}$$

Parameters	Value (UCT)
C	50.576
D_{UCT}	4.545
E_{UCT}	0.875
G_{UCT}	32.260
R_s	13.43 days

Results	Value (UCT)
RHS	32.773

3.2.3 Secondary Settling Tank Design

Calculating flux constants V_0 and n from equations (2-11),(2-12) and (2-13):

$$SSVI = DSVI \times 0.67$$

$$V_0/n = 67.9\, e^{-0.016(SSVI)}$$

$$n = 0.88 - 0.393 \log\left(V_0/n\right)$$

Parameters	Value (MLE)	Value (UCT)
$DSVI$	$120\, m\ell/g$	$120\, m\ell/g$

Results	Value (MLE)	Value (UCT)
n	$0.380\, m^3/kgTSS$	$0.380\, m^3/kgTSS$
V_0	$7.12\, m/h$	$7.12\, m/h$

Calculating maximum up-flow-velocity from equation (2-14):

$$q_{Amax} = V_0\, e^{-nXt}$$

Parameters	Value (MLE)	Value (UCT)
X_t	$5.20\, kgTSS/m^3$	$4.30\, kgTSS/m^3$

Results	Value (MLE)	Value (UCT)
q_{Amax}	$0.989\, m/h$	$1.392\, m/h$

Calculating SST area from equation (2-15):

$$A_{SST} = PWWF/q_{Amax}$$

Parameters	Value (MLE)	Value (UCT)
$PWWF$	$3200\, m^3/h$	$3200\, m^3/h$

Results	Value (MLE)	Value (UCT)
A_{SST} $(Total)$	$4\,045\ m^2$	$2874\ m^2$
Number of units	5	3
A_{SST} $(Unit)$	$809\ m^2$	$958\ m^2$
$\emptyset_{SST}$ $(Unit)$	$32.10\ m$	$34.92\ m$

3.2.4 Reactor/SST Cost Optimisation

Biological Reactor

Calculating volume of biological reactor from equation (2-16):

$$V_p = \frac{Q_{i,ADWF}S_{ti,Reactor}}{X_t}[A]$$

Parameters	Value (MLE)	Value (UCT)
$Q_{i,ADWF}$	$30\ M\ell/d$	$30\ M\ell/d$
$S_{ti,Reactor}$	$672\ mg/\ell$	$672\ mg/\ell$
$[A]$	4.12	5.99
R_s	$12.15\ days$	$13.43\ days$

Secondary Settling Tank

Calculating SST Area (A_{ST}) from equation (2-17):

$$A_{ST} = \frac{1000f_q Q_{i,ADWF}/24}{0.8V_0 exp(-nX_t)}$$

Parameters	Value (MLE)	Value (UCT)
f_q	2.56	2.56
$Q_{i,ADWF}$	$30\ M\ell/d$	$30\ M\ell/d$
n	$0.380\ m^3/kgTSS$	$0.380\ m^3/kgTSS$
V_0	$7.12\ m/h$	$7.12\ m/h$

To calculate the optimum MLSS concentration, a graph was plotted with the combined cost of the biological reactors and SSTs to find the minimum cost as can be seen in the graph below. The MLSS concentration as which the minimum combined cost is obtained is the optimum MLSS concentration.

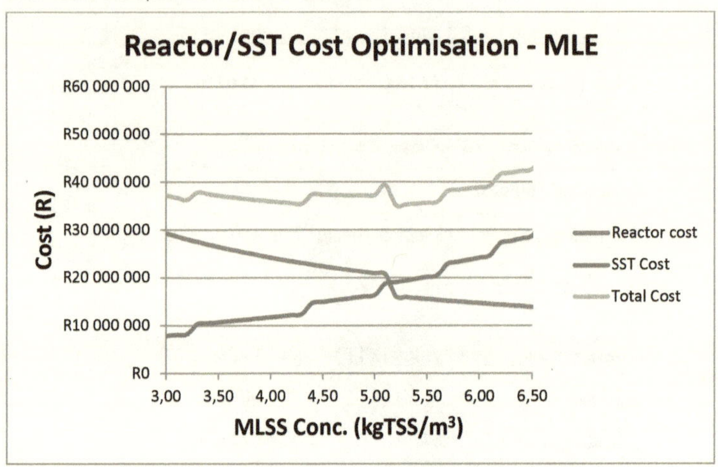

Figure 22: Reactor/SST Cost Optimisation – MLE Process Configuration

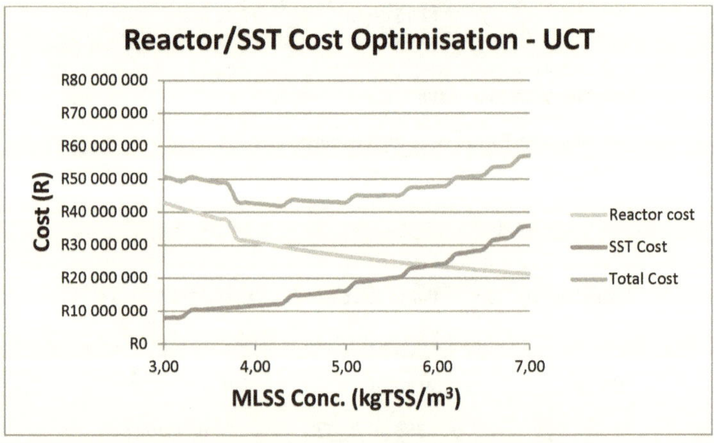

Figure 23: Reactor/SST Cost Optimisation – UCT Process Configuration

Results	Value (MLE)	Value (UCT)
X_t	$5.20\ kgTSS/m^3$	$4.30\ kgTSS/m^3$

3.2.5 Blower House Building

From Table 3 the sizing of the blower house building is as follows:

Results	Value (MLE)	Value (UCT)
Area	$240\ m^2$	$130\ m^2$

3.3 Aeration Equipment

3.3.1 Slow Speed Surface Aeration (SSSA)

Calculating R_{act} from equation (2-32):

$$R_{act} = \alpha\theta^{(T-20)}\frac{\left[\left(\frac{P_{site}-p_{site}}{P_{std}-p_{std}}\right)\left(\frac{51.6}{31.6+T}\right)\beta C_s - C_L\right]}{C_s} \times R_{std}$$

Parameters	Value (MLE)	Value (UCT)
R_{std}	$2.00\ kgO/kWh$	$2.00\ kgO/kWh$
α	0.80	0.80
β	0.95	0.95
θ	1.024	1.024
C_s	$9.07\ mgO/\ell$	$9.07\ mgO/\ell$
C_L	$2.00\ mgO/\ell$	$2.00\ mgO/\ell$
$Altitude$	$40\ m$	$40\ m$
T	$22°C$	$22°C$

Adjusting pressure (P) for increase in attitude (equation (2-20)):

$$P = 10^{2.88117-0.000053407\times Alt)}$$

$$P = 756.892\ mmHg$$

Adjusting Saturated Water Vapour Pressure for increase in temperature (equation (2-21)):

$$p = 17.51(1.0639)^{(T-20)}$$

$$p = 19.819$$

Results	Value (MLE)	Value (UCT)
R_{act}	1.153 kgO/kWh	1.153 kgO/kWh

From equation (2-33) the power requirement can be calculated:

$$P_w = R_{act} \times \text{OUR} \times V_{Aer}$$

Results	Value (MLE)	Value (UCT)
P_w	481.00 kW	456.00 kW

This power requirement is used for the calculation of the operational cost.

The aerator equipment sizes chosen is as follows:

Aerator size	Value (MLE)	Value (UCT)
80 kW	2	2
65 kW	2	2
55 kW	4	4
Total power	510 kW	510 kW

This power requirement is used for the calculation of the capital cost.

3.3.2 Fine Bubble Diffused Aeration

3.3.2.1 Airflow Calculation

Calculating $\frac{OTR_{site}}{OTR_{std}}$ from equation (2-34):

$$\frac{OTR_{site}}{OTR_{std}} = \left\{ \alpha\theta^{T-20} \frac{\left[\left(\frac{P_{site} - p_{site}}{P_{std} - p_{std}}\right)\left(\frac{P_{site} + 73.53hf - P_{site}}{P_{site} - p_{site}}\right)\left(\frac{51.6}{31.6 + T}\right)\beta C_{Sstd}F - C_L\right]}{C_s} \right\} = \{A\}$$

Parameters	Value (MLE)	Value (UCT)
α	0.80	0.80
β	0.95	0.95
θ	1.024	1.024
F	0.90	0.90
f	0.325	0.325
C_s	$9.07\ mgO/\ell$	$9.07\ mgO/\ell$
C_L	$2.00\ mgO/\ell$	$2.00\ mgO/\ell$
Altitude	$40\ m$	$40\ m$
T_{Site}	$22°C$	$22°C$
T_{Blower}	$36°C$	$36°C$

Adjusting pressure (P) for increase in attitude (equation (2-20)):

$$P = 10^{2.88117 - 0.000053407 \times Alt)}$$

$$P = 756.892\ mmHg$$

Adjusting Saturated Water Vapour Pressure for increase in temperature (equation (2-21)):

$$p = 17.51(1.0639)^{(T-20)}$$

$$p = 19.819$$

$$\frac{OTR_{site}}{OTR_{std}} = 0.372 = \{A\}$$

From equation (2-35)

$$OSR_{std} = Con_{O2,std} \times Q_{Air,std} = {OTR_{std}}/{SOTE_{std}}$$

and equation (2-37)

$$OTR_{site} = AOR$$

Parameters	Value (MLE)	Value (UCT)
$Con_{O2,std}$	$0.298\ kgO/m^3$	$0.298\ kgO/m^3$
$OTE\ \%\ /\ m$	$7.00\ \%$	$7.00\ \%$
Submergence depth	$4.25\ m$	$4.25\ m$
AOR_{site}	$6\ 256\ kgO/h$	$5939\ kgO/h$

Calculate $SOTE_{std}$ from equation (2-36)

$$SOTE_{std} = OTE \times submergence\ depth$$

$$SOTE_{std} = 0.2975$$

From equation (2-38)

$$AOR_{site} = OTR_{std}\{A\} = Con_{O2,std} \times Q_{Air,std} \times SOTE_{std}\{A\}$$

With this, $Q_{Air,std}$ can be solved (equation (2-39))

$$Q_{Air,std} = {AOR_{site}}/{(Con_{O2,std} \times SOTE_{std}\{A\})}$$

Results	Value (MLE)	Value (UCT)
$Q_{Air,std}$	$18\ 043\ m^3/h$	$17\ 127\ m^3/h$

Calculate $Q_{Air,Blower}$ at blower temperature from equation (2-40):

$$Q_{Air,Blower} = {AOR_{site}}/{(Con_{O2,Blower} \times SOTE_{std}\{A\})}$$

Results	Value (MLE)	Value (UCT)
$Q_{Air,Blower}$	$19\ 112\ m^3/h$	$18\ 142\ m^3/h$

Calculate the number of the diffusers from the equation (2-47):

$$No.of\ diffusers = {Area_{Aer} \times DD}/{Area_{diff}}$$

Parameters	Value (MLE)	Value (UCT)
$Area_{Aer}$	$2\,293\ m^2$	$3\,333\ m^2$
DD	13.85 %	9.05 %
$Area_{diff}$	$0.0415\ m^2$	$0.0415\ m^2$

Results	Value (MLE)	Value (UCT)
No. of diffusers	7 645	7 257

3.3.2.2 Pressure Rating Calculation

Calculating the pressure rating of Blower.

Headloss through aeration pipes:

Parameters	Value (MLE)	Value (UCT)
Pipe length	$100\ m$	$100\ m$
Pressure drop	$0.75\ kPa/30.5m$	$0.75\ kPa/30.5m$

$Headloss\ (pipes) = 2.46 + 0.5 = 2.96\ kPa$

Aeration pipe size:

Parameters	Value (MLE)	Value (UCT)
Airflow	$19\,112\ m^3/h$	$18\,142\ m^3/h$

From Table 6:

Results	Value (MLE)	Value (UCT)
Pipe size	$500\ mm$	$500\ mm$

Headloss through diffusers:

Parameters	Value (MLE)	Value (UCT)
Diffuser type	SSI ECD270 Disc Diffuser	SSI ECD270 Disc Diffuser
Airflow / diffuser	$2.50\ m^3/h$	$2.50\ m^3/h$

From equation (2-53):

$$Headloss = 33.19(Airflow\ rate) + 171.81$$

Results	Value (MLE)	Value (UCT)
Headloss (diffusers)	$2.50\ kPa$	$2.50\ kPa$

Submergence of diffusers:

Parameters	Value (MLE)	Value (UCT)
Submergence depth	$4.25\ m$	$4.25\ m$

From equation (2-54):

$$Head = 9.81 \times depth\ of\ submergence$$

Results	Value (MLE)	Value (UCT)
Head	$41.69\ kPa$	$41.69\ kPa$

Fouling of diffusers:

$$Headloss\ (fouling) = 0.62\ kPa$$

Total headloss through FBDA system:

Results	Value (MLE)	Value (UCT)
Headloss (total)	$47.78\ kPa = 0.47\ atm$	$47.78\ kPa = 0.47\ atm$

3.3.2.3 Blower Selection

Calculating blower power requirements (P_w) from equation (2-55):

$$P_w = \frac{wRT_1}{29.7\ n\ e}\left[\left(\frac{p_2}{p_1}\right)^{0.283} - 1\right]$$

Parameters	Value (MLE)	Value (UCT)
R	$8.314\ J/(mol.\,K)$	$8.314\ J/(mol.\,K)$
T_1	$319.15\ K$	$319.15\ K$
n	0.283	0.283
e	0.80	0.80

Calculating pressure at inlet of blower (p_1) for increase in attitude and temperature (equation (2-56)):

$$p_1 = p_0 \exp\left(-\frac{Mg}{RT}h\right)$$

$$p_1 = 100\,878.15\ Pa = 0.9956\ atm$$

To calculate the outlet pressure of blower (p_2) the headlosses calculated above must be added to the inlet pressure.

Parameters	Value (MLE)	Value (UCT)
Headloss (total)	0.47 atm	0.47 atm

Results	Value (MLE)	Value (UCT)
p_2	1.47 atm	1.47 atm

Calculating the weight flow of air from equation (2-57):

$$w = Q_{Air,Blower} \times \rho$$

Parameters	Value (MLE)	Value (UCT)
$Q_{Air,Blower}$	19 112 m^3/h	18 142 m^3/h

To density of air for increase in altitude and temperature (equation (2-60))

$$\rho_b = \frac{P_a \exp\left[-\dfrac{gM(Z_b - Z_a)}{RT_a}\right]M}{RT_b}$$

Parameters	Value (MLE)	Value (UCT)
P_a	101 325 Pa	101 325 Pa
g	9.8107	9.8107
M	28.97	28.97
$Z_b - Z_a$	40 m	40 m
R	8.314 $J/(mol.K)$	8.314 $J/(mol.K)$
T_a	273.15 K	273.15 K
T_b	319.15 K	319.15 K

Results	Value (MLE)	Value (UCT)
ρ_b	$1.101\ kg/m^3$	$1.101\ kg/m^3$

Results	Value (MLE)	Value (UCT)
w	$5.84\ m^3/s$	$5.55\ m^3/s$

Results	Value (MLE)	Value (UCT)
P_w	$268\ kW$	$254\ kW$

Incorporating the 15% energy saving of the aeration control system:

$$P_{w,act} = P_w \times (100\% - 15\%)$$

Results	Value (MLE)	Value (UCT)
$P_{w,act}$	$228\ kW$	$216\ kW$

This power requirement is used for the calculation of the operational cost. The blower sizes chosen is as follows:

Blower size	Value (MLE)	Value (UCT)
200 kW	3 (2 duty / 1 standby)	3 (2 duty / 1 standby)
Total power	600 kW	480 kW

This power requirement is used for the calculation of the capital cost.

3.4 Cost Modelling

3.4.1 Capital Cost

3.4.1.1 Biological Reactors (SSSA and FBDA)

Parameters used:

Infrastructure	Cost Function	Cost (MLE)	Cost (UCT)
Biological Reactor (SSSA)	$30.08V^{0.656} \times 1000$	R 1 638 2364	R 31 467 184
Biological Reactor (FBDA)	$26.35V^{0.663} \times 1000$	R 16 121 807	R 29 469 284
Secondary Settling Tanks (SSSA)	$137.67\emptyset^{0.957} \times 1000$	R 20 471 296	R 12 383 777
Secondary Settling Tanks (FBDA)	$137.67\emptyset^{0.957} \times 1000$	R 19 032 560	R 12 383 777
Blower House Building	$85.12(Area)^{0.514} \times 1000$	R 1 483 638	R 1 038 959
Slow Speed Surface Aeration–Surface Aerators	$31.89(kW)^{0.687} \times 1000$	R 2 310 762	R 2 310 762
SSSA–Electrical Equipment		R 346 614	R 346 614
Fine Bubble Diffused Aeration–Blower System	$886.53(kW)^{0.345} \times 1000$	R 8 422 878	R 7 825 912
Main Aeration Header	$2322.5(Dia)^{0.802}$	R 366 207	R 366 207
Fine Bubble Diffused Aeration–Diffuser System	$1173.8(SOR) + 260769$	R 2 445 362	R 2 334 494
FBDA– Electrical Equipment		R 631 716	R 586 943

3.4.1.2 Total Capital Cost

Results	Value (MLE)	Value (UCT)
Slow Speed Surface Aeration	R 39 511 028	R 46 508 337
Fine Bubble Diffused Aeration	R 48 078 160	R 53 639 370

3.4.2 Operational Cost

Calculating the electricity cost based on section 2.8.2 and power requirement calculated in section 3.3 above:

3.4.2.1 Slow Speed Surface Aeration

Parameters used:

Parameters	Value (MLE)	Value (UCT)
Power	481 kW	456 kW
Running time / year	365 days	365 days
Energy charge / kW / year (high demand)	R2 353.50	R2 353.50
Energy charge / kW / year (low demand)	R4 407.34	R4 407.34
Transmission network charge	R 11.20 / kVA / month	R 11.20 / kVA / month
Service charge	R220.24 / day	R220.24 / day
Admin charge	R 99.28 / day	R 99.28 / day

3.4.2.2 Fine Bubble Diffused Aeration

Parameters used:

Parameter	Value (MLE)	Value (UCT)
Power kW	228 kW	216 kW
Running time / year	365 days	365 days
Energy charge / kW / year (high demand)	R2 353.50	R2 353.50
Energy charge / kW / year (low demand)	R4 407.34	R4 407.34
Transmission network charge	R 11.20 / kVA / month	R 11.20 / kVA / month
Service charge	R220.24 / day	R220.24 / day
Admin charge	R 99.28 / day	R 99.28 / day

3.4.2.3 Total Operational Cost

Results	Value (MLE)	Value (UCT)
Slow Speed Surface Aeration	R 3 332 773	R 3 159 552
Fine Bubble Diffused Aeration	R 1 574 952	R 1 495 023

3.4.3 Maintenance Cost

Calculating maintenance cost from Table 11:

3.4.3.1 Slow Speed Surface Aeration

	Value (MLE)		Value (UCT)	
Results	Civil	Mech & Elec	Civil	Mech & Elec
Biological Reactors	R 163 824		R 314 672	
Secondary Settling Tanks	R 204 713		R 123 838	
Surface Aerators		R 106 295		R 106 295

3.4.3.2 Fine Bubble Diffused Aeration

	Value (MLE)		Value (UCT)	
Results	Civil	Mech & Elec	Civil	Mech & Elec
Biological Reactors	R 161 218		R 294 693	
Secondary Settling Tanks	R 190 326		R 123 838	
Blower House Building	R 14 836		R 10 390	
Blower System		R 362 184		R 336 514
Diffuser System		R 97 815		R 93 380

3.4.3.3 Total Maintenance Cost

Results	Value (MLE)	Value (UCT)
Slow Speed Surface Aeration	R 474 832	R 544 805
Fine Bubble Diffused Aeration	R 825 780	R 858 814

3.4.4 Comparison of Total Cost

3.4.4.1 Slow Speed Surface Aeration

Results	Value (MLE)	Value (UCT)
Capital Cost	R 39 511 028	R 46 508 337
Operational Cost per year	R 3 332 773	R 3 159 552
Maintenance Cost per year	R 474 832	R 544 805

Calculating Net Present Value from equation (2-65):

$$NPV = \sum_{t=1}^{T} \frac{C_t}{(1+r)^t} - C_0$$

MLE Process Configuration:

Year	Interest and redemption on Capital Cost	Maintenance (civil) escalated at 7.00% p.a.	Maintenance (mech & elec) escalated at 7.00% p.a.	Electrical Operating escalated at 10.00% p.a.	Total: De-escalation at % 7.00 p.a.
1	R 4 800 324	R 368 537	R 106 295	R 3 666 051	R 84 101 295
2	R 4 800 324	R 394 334	R 113 736	R 4 032 656	R 89 988 386
3	R 4 800 324	R 421 937	R 121 697	R 4 435 921	R 96 287 573
4	R 4 800 324	R 451 473	R 130 216	R 4 879 513	R 103 027 703
5	R 4 800 324	R 483 076	R 139 331	R 5 367 465	R 110 239 642
6	R 4 800 324	R 516 892	R 149 084	R 5 904 211	R 117 956 417
7	R 4 800 324	R 553 074	R 159 520	R 6 494 632	R 126 213 366
8	R 4 800 324	R 591 789	R 170 687	R 7 144 095	R 135 048 302
9	R 4 800 324	R 633 214	R 182 635	R 7 858 505	R 144 501 683
10	R 4 800 324	R 677 539	R 195 419	R 8 644 356	R 154 616 801
11	R 4 800 324	R 724 967	R 209 098	R 9 508 791	R 165 439 977
12	R 4 800 324	R 775 715	R 223 735	R 10 459 670	R 177 020 775
13	R 4 800 324	R 830 015	R 239 397	R 11 505 637	R 189 412 229
14	R 4 800 324	R 888 116	R 256 155	R 12 656 201	R 202 671 085
15	R 4 800 324	R 950 284	R 274 085	R 13 921 821	R 216 858 061
16	R 4 800 324	R 1 016 804	R 293 271	R 15 314 003	R 232 038 126
17	R 4 800 324	R 1 087 980	R 313 800	R 16 845 403	R 248 280 795
18	R 4 800 324	R 1 164 139	R 335 766	R 18 529 944	R 265 660 450
19	R 4 800 324	R 1 245 628	R 359 270	R 20 382 938	R 284 256 682
20	R 4 800 324	R 1 332 822	R 384 419	R 22 421 232	R 304 154 649
Tot	R 96 006 475	R 15 108 336	R 4 357 618	R 209 973 046	R 325 445 475

UCT Process Configuration:

Year	Interest and redemption on Capital Cost	Maintenance (civil) escalated at 7.00% p.a.	Maintenance (mech & elec) escalated at 7.00% p.a.	Electrical Operating escalated at 10.00% p.a.		Total: De-escalation at % 7.00 p.a.
1	R 5 650 450	R 438 510	R 106 295	R 3 475 507		R 86 416 146
2	R 5 650 450	R 469 205	R 113 736	R 3 823 058		R 92 465 276
3	R 5 650 450	R 502 050	R 121 697	R 4 205 364		R 98 937 845
4	R 5 650 450	R 537 193	R 130 216	R 4 625 900		R 105 863 494
5	R 5 650 450	R 574 797	R 139 331	R 5 088 490		R 113 273 939
6	R 5 650 450	R 615 032	R 149 084	R 5 597 339		R 121 203 115
7	R 5 650 450	R 658 085	R 159 520	R 6 157 073		R 129 687 333
8	R 5 650 450	R 704 151	R 170 687	R 6 772 781		R 138 765 446
9	R 5 650 450	R 753 441	R 182 635	R 7 450 059		R 148 479 027
10	R 5 650 450	R 806 182	R 195 419	R 8 195 065		R 158 872 559
11	R 5 650 450	R 862 615	R 209 098	R 9 014 571		R 169 993 638
12	R 5 650 450	R 922 998	R 223 735	R 9 916 028		R 181 893 193
13	R 5 650 450	R 987 608	R 239 397	R 10 907 631		R 194 625 717
14	R 5 650 450	R 1 056 740	R 256 155	R 11 998 394		R 208 249 517
15	R 5 650 450	R 1 130 712	R 274 085	R 13 198 234		R 222 826 983
16	R 5 650 450	R 1 209 862	R 293 271	R 14 518 057		R 238 424 872
17	R 5 650 450	R 1 294 552	R 313 800	R 15 969 863		R 255 114 613
18	R 5 650 450	R 1 385 171	R 335 766	R 17 566 849		R 272 972 636
19	R 5 650 450	R 1 482 133	R 359 270	R 19 323 534		R 292 080 720
20	R 5 650 450	R 1 585 882	R 384 419	R 21 255 887		R 312 526 371
Tot	R 113 008 995	R 17 976 917	R 4 357 618	R 199 059 686		R 334 403 217

94

3.4.4.2 Fine Bubble Diffused Aeration

Results	Value (MLE)	Value (UCT)
Capital Cost	R 48 078 160	R 53 639 370
Operational Cost per year	R 1 574 952	R 1 495 023
Maintenance Cost per year	R 825 780	R 858 814

Calculating Net Present Value from equation (2-65):

$$NPV = \sum_{t=1}^{T} \frac{C_t}{(1+r)^t} - C_0$$

MLE Process Configuration:

Year	Interest and redemption on Capital Cost	Maintenance (civil) escalated at 7.00% p.a.	Maintenance (mech & elec) escalated at 7.00% p.a.	Electrical Operating escalated at 10.00% p.a.	Replacement of diffusers escalated at 7.00% p.a.	Total: De-escalation at % 7.00 p.a.
1	R 5 841 173	R 365 782	R 459 998	R 1 732 447		R 65 548 054
2	R 5 841 173	R 391 387	R 492 198	R 1 905 692		R 70 136 417
3	R 5 841 173	R 395 301	R 526 652	R 2 096 261		R 75 045 967
4	R 5 841 173	R 399 254	R 563 518	R 2 305 887		R 80 299 184
5	R 5 841 173	R 403 246	R 602 964	R 2 536 476		R 85 920 127
6	R 5 841 173	R 407 279	R 645 171	R 2 790 124		R 91 934 536
7	R 5 841 173	R 411 351	R 690 333	R 3 069 136	R 3 926 718	R 98 369 954
8	R 5 841 173	R 415 465	R 738 657	R 3 376 050		R 105 255 850
9	R 5 841 173	R 419 620	R 790 363	R 3 713 655		R 112 623 760
10	R 5 841 173	R 423 816	R 845 688	R 4 085 020		R 120 507 423
11	R 5 841 173	R 428 054	R 904 886	R 4 493 522		R 128 942 943
12	R 5 841 173	R 432 335	R 968 228	R 4 942 874		R 137 968 949
13	R 5 841 173	R 436 658	R 1 036 004	R 5 437 162		R 147 626 775
14	R 5 841 173	R 441 024	R 1 108 524	R 5 980 878	R 6 305 450	R 157 960 649
15	R 5 841 173	R 445 435	R 1 186 121	R 6 578 966		R 169 017 895
16	R 5 841 173	R 449 889	R 1 269 150	R 7 236 862		R 180 849 147
17	R 5 841 173	R 454 388	R 1 357 990	R 7 960 548		R 193 508 588
18	R 5 841 173	R 458 932	R 1 453 049	R 8 756 603		R 207 054 189
19	R 5 841 173	R 463 521	R 1 554 763	R 9 632 264		R 221 547 982
20	R 5 841 173	R 468 156	R 1 663 596	R 10 595 490		R 237 056 341
Tot	R 116 823 453	R 8 510 891	R 18 857 854	R 99 225 917	R 10 232 168	R 253 650 284

UCT Process Configuration:

Year	Interest and redemption on Capital Cost	Maintenance (civil) escalated at 7.00% p.a.	Maintenance (mech & elec) escalated at 7.00% p.a.	Electrical Operating escalated at 10.00% p.a.	Replacement of diffusers escalated at 7.00% p.a.	Total: De-escalation at % 7.00 p.a.
1	R 6 516 822	R 428 920	R 429 894	R 1 644 525		R 67 679 571
2	R 6 516 822	R 458 945	R 459 987	R 1 808 978		R 72 417 141
3	R 6 516 822	R 463 534	R 492 186	R 1 989 876		R 77 486 341
4	R 6 516 822	R 468 169	R 526 639	R 2 188 863		R 82 910 385
5	R 6 516 822	R 472 851	R 563 503	R 2 407 750		R 88 714 112
6	R 6 516 822	R 477 580	R 602 949	R 2 648 525		R 94 924 100
7	R 6 516 822	R 482 355	R 645 155	R 2 913 377	R 3 748 687	R 101 568 787
8	R 6 516 822	R 487 179	R 690 316	R 3 204 715		R 108 678 602
9	R 6 516 822	R 492 051	R 738 638	R 3 525 186		R 116 286 104
10	R 6 516 822	R 496 971	R 790 343	R 3 877 705		R 124 426 131
11	R 6 516 822	R 501 941	R 845 667	R 4 265 476		R 133 135 960
12	R 6 516 822	R 506 960	R 904 863	R 4 692 023		R 142 455 478
13	R 6 516 822	R 512 030	R 968 204	R 5 161 225		R 152 427 361
14	R 6 516 822	R 517 150	R 1 035 978	R 5 677 348	R 6 019 573	R 163 097 276
15	R 6 516 822	R 522 322	R 1 108 496	R 6 245 083		R 174 514 086
16	R 6 516 822	R 527 545	R 1 186 091	R 6 869 591		R 186 730 072
17	R 6 516 822	R 532 820	R 1 269 117	R 7 556 550		R 199 801 177
18	R 6 516 822	R 538 149	R 1 357 956	R 8 312 205		R 213 787 259
19	R 6 516 822	R 543 530	R 1 453 013	R 9 143 426		R 228 752 367
20	R 6 516 822	R 548 965	R 1 554 723	R 10 057 768		R 244 765 033
Tot	R 130 336 444	R 9 979 968	R 17 623 716	R 94 190 196	R 9 768 260	R 261 898 585

Net Present Valuse	Value (MLE)	Value (UCT)
Slow Speed Surface Aeration	R 84 101 295	R 86 416 146
Fine Bubble Diffused Aeration	R 65 548 054	R 67 679 571

For the analysis and discussions of the Kelvin Jones data and results refer to section 4.4 below.

3.5 Generalised Model

As explained and calculated above, the Kelvin Jones data was chosen as the sample project for this investigation. However, in order for the design model to be used for varies inflows and wastewater characteristics, a more generalised model with varying data groups were set up. As the proposed upgraded modules for the Kelvin Jones WWTP will have a maximum capacity of 30 Mℓ/d and it is unlikely that a new or upgraded plant will exceed this capacity without phasing the project, the inflow rate for the data groups were capped at 30 Mℓ/d. As the main objective of the study is to find the VTP, the inflow rate was increased in increments of 2.5 Mℓ/d, starting at 2.5 Mℓ/d and ending at 30 Mℓ/d. For the benchmark data group, constant inflow wastewater characteristics were used and only the flow was varied. For this data group the inflow COD concentration were chosen as 750 mg/ℓ and the other inflow characteristics were calculated to represent typical values based on this COD concentration.

In order to see the effect different COD concentrations will have on the outcome, similar data groups were also created with different COD concentrations. Details of the generalised inflow data groups that were used, are presented in Appendix A.

To test the financial viability between the slow speed surface aeration and the fine bubble diffused aeration technology, the capital, operational and maintenance costs between these two systems were calculated (similar to the calculation example above) using the different inflow data groups.

4. RESULTS AND DISCUSSION

The design methodology and cost modelling as explained in section 2 was used to set up a excel based plant design model. The generalised data (as represented in Appendix A) was used as the input for the model. As mentioned, the MLE and UCT process configurations were used and both raw and settled wastewater were investigated.

As the main objective of the study is to test the financial viability between slow speed surface aeration (SSSA) and fine bubble diffused aeration (FBDA) i.e. find the VTP, the capital, operational and maintenance costs between these two systems were calculated.

4.1 Benchmark Results

As mentioned in section 3.5, to create a benchmark data group, an inflow COD concentration of 750 mg/ℓ were chosen and the other inflow characteristics were calculated to represent typical values based on this COD concentration. Based on these inflow characteristics, incorporating the varying inflow rate (2.5 Mℓ/d to 30 Mℓ/d) and taking into account different process configurations (MLE and UCT), the total cost of the two different aeration technologies were calculated.

From these benchmark results it was concluded that the VTP is reached at much lower flow rate than originally estimated. For this reason, the varying inflow rates were lowered to focus on a smaller area, i.e. 1.0 Mℓ/d to 12.0 Mℓ/d, to improve the accuracy of the results.

4.1.1 MLE Process Configuration

The graph below represents the Total Cost vs Inflow Rate (Mℓ/d) for both aeration technologies using the MLE process configuration for the Raw Wastewater.

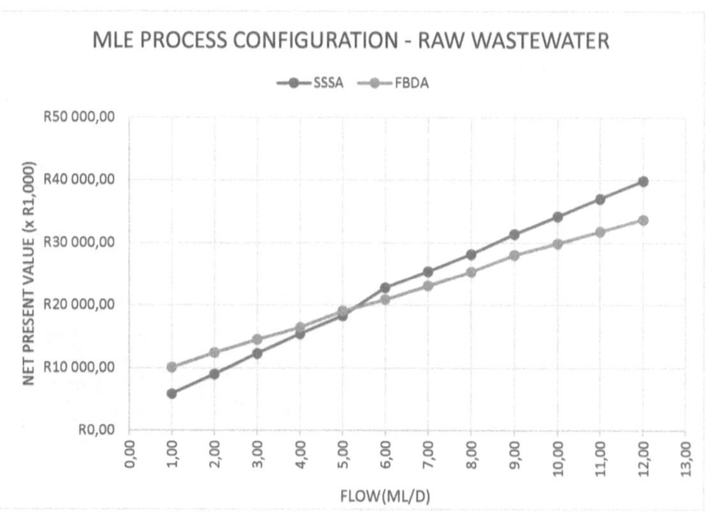

Figure 24: Total Cost vs Inflow Rate – MLE process configuration – Raw wastewater

From the graph it is evident that the VTP is reached at an inflow rate of 5.30 Mℓ/d. For this reason, when designing a WWTP based on the MLE process configuration and similar inflow characteristics, designers and investors should start to consider the FBDA aeration option for any plant with an inflow rate of more than 5.30 Mℓ/d.

A similar graph, representing Total Cost vs Inflow Rate (Mℓ/d), was plotted for the Settled Wastewater (MLE process configuration) and is presented below.

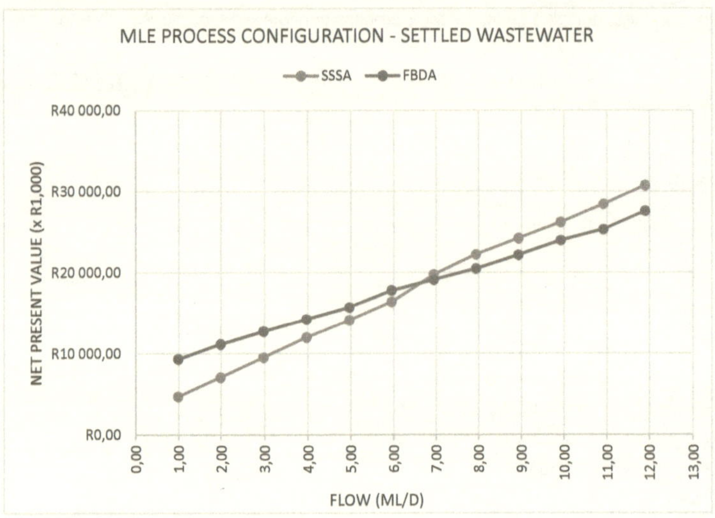

Figure 25: Total Cost vs Inflow Rate – MLE process configuration – Settled wastewater

For the Settled wastewater the VTP is reached at an inflow rate of 6.60 Mℓ/d. The FBDA aeration option will therefore be the more viable option for any plant with an inflow rate of more than 6.60 Mℓ/d.

It is interesting to note is that the FBDA aeration option becomes the more viable option at a higher inflow rate for Settled Wastewater than for Raw Wastewater. The main reason for this is that less oxygen (or energy) is required to treat Settled Wastewater, as the concentration of certain constituents (mainly COD) are lowered during primary sedimentation. The effect of the more energy efficient FBDA system are therefore less profound for Settled Wastewater.

4.1.2 UCT Process Configuration

Similar to the graphs done for the MLE process configuration, the Total Cost vs Inflow Rate (Mℓ/d) graphs were also completed for the UCT process configuration.

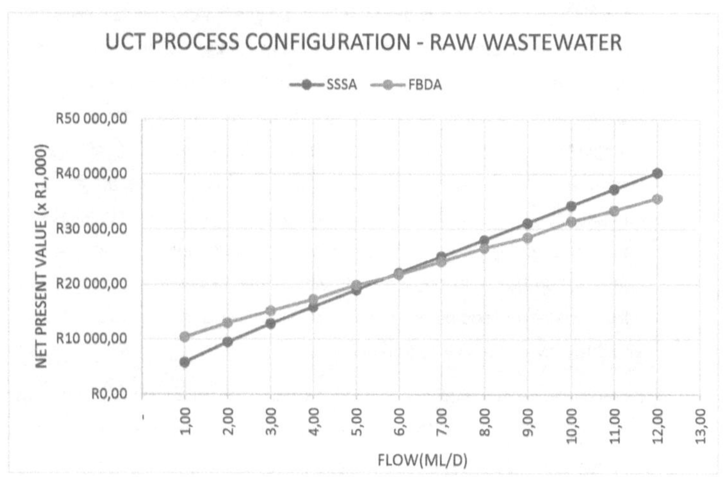

Figure 26: Total Cost vs Inflow Rate – UCT process configuration – Raw wastewater

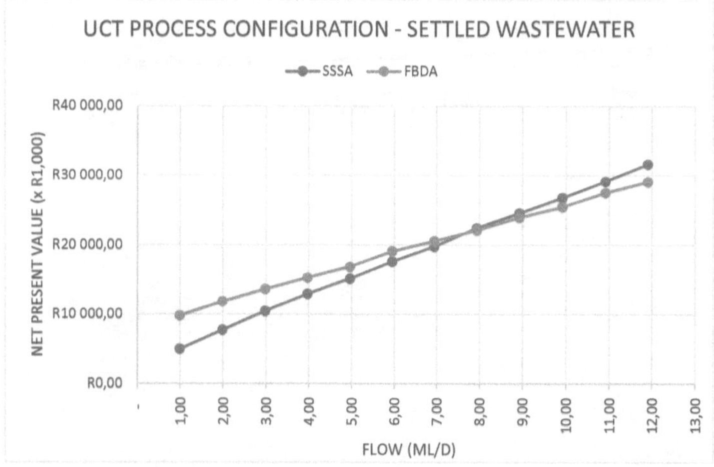

Figure 27: Total Cost vs Inflow Rate – UCT process configuration – Settled wastewater

From these graphs it is evident that the VTP is reached at an inflow rate of 5.80 Mℓ/d, when looking at Raw Wastewater and at an inflow rate of 7.60 Mℓ/d when looking at Settled Wastewater.

Again, it can be noted that the VTP is at a higher inflow rate for Settled Wastewater than for Raw Wastewater (same reasoning as for the MLE process configuration). It is also interesting to note that the VTP of both the Raw and Settled Wastewater are slightly higher in the UCT process configuration than for the MLE process configuration. The reason for this is that the UCT process configuration can treat slightly less wastewater with the same size infrastructure as the MLE process configuration can. As the inflow rates and wastewater characteristics remained the same for both the UCT and MLE process configurations, it means the UCT process configuration has a slightly longer balanced sludge age and therefore slightly larger infrastructure and equipment is required. This means the capital cost is slightly higher and this moves the VTP to the right.

It can therefore be concluded that incorporating the MLE process configuration will be cheaper than the UCT process configuration. The MLE process configuration is however not capable of biological removal of phosphorous and should phosphorous removal be required as part of the effluent standard, the UCT process configuration should be chosen. Chemical phosphorous removal by means of Ferric Chloride (or a similar chemical) is however an option when the MLE process configuration is chosen. The chemicals are however extremely expensive and therefore chemical phosphorous removal is only a viable option for small WWTPs.

4.2 Effect of COD Concentration

To incorporate the effect that a higher or lower COD concentration (and the other inflow characteristics calculated to represent typical values based on the COD concentration) will have on the VTP, data groups with different COD concentrations were introduced as summarised below (for more detail see Appendix A):

Dataset	Raw Wastewater COD Concentration	Settled Wastewater COD Concentration
Data Group 1	600	360
Data Group (Benchmark)	750	450
Data Group 3	900	540
Data Group 4	1050	630
Data Group 5	1200	720

Table 13: COD Concentrations of Data Groups

As with the benchmark results, the total cost of the two different aeration technologies for each data group were calculated. For each data group the VTP was calculated for both process configuration and for Raw and Settled Wastewater.

4.2.1 MLE Process Configuration

The graphs below represent the VTP for varying COD concentrations, based on the MLE process configuration for Raw Wastewater and Settled Wastewater.

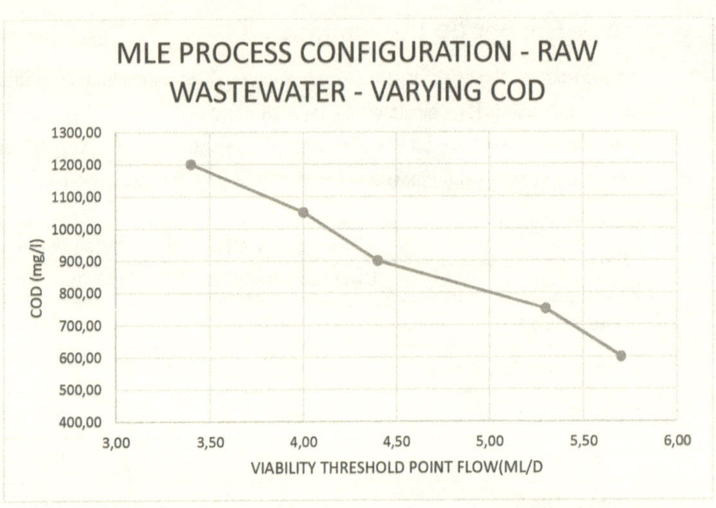

Figure 28:Viability Threshold Points for varying COD concentrations - MLE process configuration – Raw wastewater

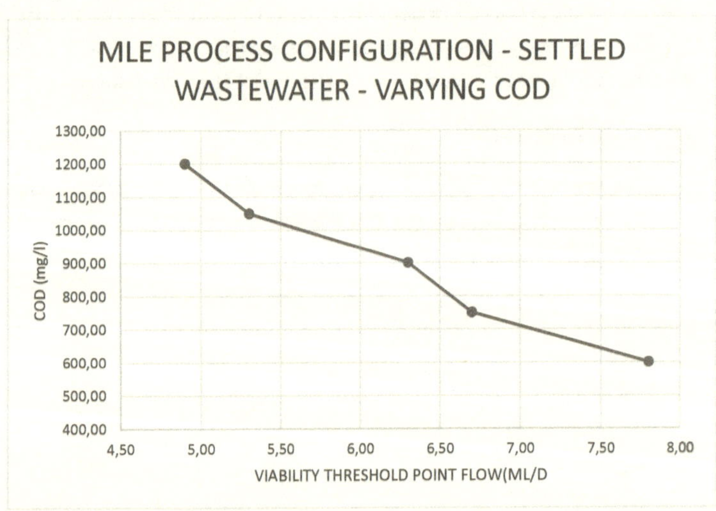

Figure 29: Viability Threshold Points for varying COD concentrations - MLE process configuration – Settled wastewater

From the graphs it is evident that an increase in COD concentration, will decrease the inflow rate at which the VTP is reached. The reason is that an increase in COD will increase the energy demand of the aeration and therefore make the effect of the more energy efficient FBDA system more profound. It can therefore be concluded that for WWTPs receiving influent with a high COD concentration, the more efficient FBDA technology will almost certainly be the most viable option. Therefore, when designing WWTP's with a high industrial influent component, designers should definitely consider the FDBA technology.

4.2.2 UCT Process Configuration

Similar to the graphs done for the MLE process configuration, the VTP for varying COD concentrations graphs were also completed for the UCT process configuration.

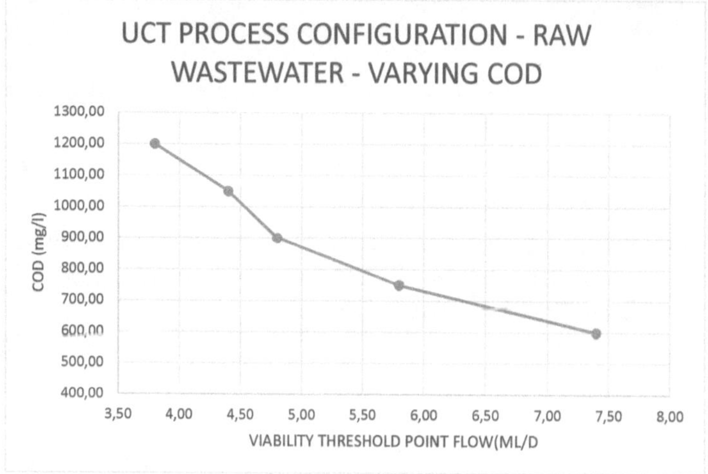

Figure 30: Viability Threshold Points for varying COD concentrations - UCT process configuration – Raw wastewater

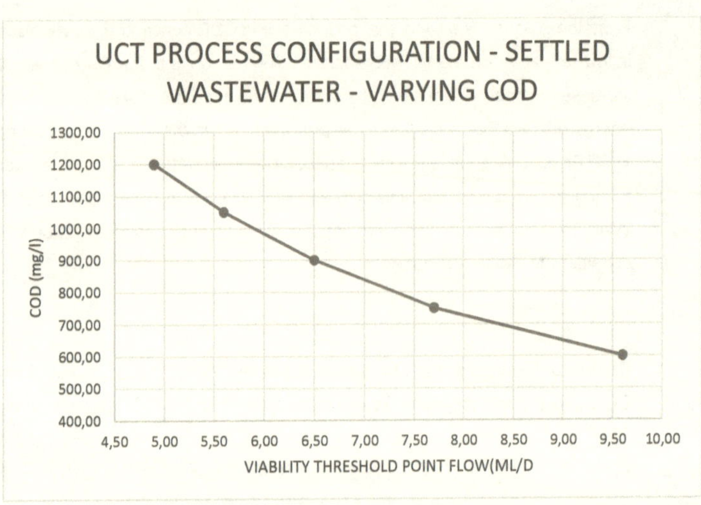

Figure 31: Viability Threshold Points for varying COD concentrations - UCT process configuration – Settled wastewater

Similar to the MLE results, the graphs above also show that with an increase in COD concentration, the inflow rate at which the VTP is reached will decrease.

From these graphs, designers and investors can therefore get the inflow rate at which the viability threshold point is reached for multiple COD concentration when the MLE and UCT process configurations are chosen.

4.3 Results Summary

In order to make it easier for designers to interpret the results above, Table 14 below
gives a summary of the Viability Threshold Points for the different input data:

	Viability Threshold Points (Mℓ/d)			
	MLE Process Configuration		UCT Process Configuration	
Data Group	Raw WW	Settled WW	Raw WW	Settled WW
Data Group 1	5.70	7.80	7.40	9.60
Data Group 2 (Benchmark)	5.30	6.60	5.80	7.60
Data Group 3	4.40	6.30	4.80	6.50
Data Group 4	4.00	5.30	4.40	5.60
Data Group 5	3.40	4.90	3.80	4.90

Table 14: Summary of Viability Threshold Points for different input data

With the information in the table above, together with the input data represented in
Appendix A, it is possible guide designers and investors to choose the correct aeration
technology given specific inflow data.

4.4 Kelvin Jones WWTP

From the Kelvin Jones WWTP results represented in section 3 it is evident that the FBDA technology would be the most viable option for upgrade of the Kelvin Jones WWTP. It is also possible to predict at what flow the VTP will be reached for the Kelvin Jones WWTP. As the Kelvin Jones WWTP is operated in a UCT process configuration, only the UCT results are applicable. With the inflow COD concentration of the Kelvin Jones WWTP measured at 672 mg/ℓ for Raw wastewater and 398 mg/ℓ for Settled wastewater, the Kelvin Jones data falls about halfway between Data Group 1 and Data Group 2. Averaging the VTPs between Data Group 1 and Data Group 2 gives a VTP of 5.50 Mℓ/d for Raw wastewater and 6.60 Mℓ/d for Settled wastewater. The Kelvin Jones WWTP is however a special case as it is an existing plant that will be upgraded and the existing plant already utilises slow speed surface aeration. The capital cost to convert the existing plant to a FBDA system will therefore be more expensive than keeping the plant on the SSSA system and the VTP predicted by the results above might need adjustment. The designed flow is however so much higher than the VTP predicted that it can still be safely assumed that the FBDA technology would be the most viable option.

4.5 Sensitivity Analysis of Key Technical Design Parameters

In order to portray the dependence and sensitivity of the chosen or stipulated values of the key technical design parameters, a sensitivity analysis was done on these parameters. The following parameters were investigated:

- Conversion factor alpha (α);
- Conversion factor beta (β);
- Surface aeration standard oxygen transfer rate (R_{std});
- Dissolved oxygen control point;
- Standard oxygen transfer efficiency (SOTE);
- Maximum water temperature;

The "Benchmark" data group (Raw Wastewater) was chosen, and the value of each parameter was decreased by 10% and the effect on the VTP's were calculated. The following table give the results of this sensitivity analysis in descending order.

Parameter	Percentage change on VTP's	
	MLE	UCT
Maximum water temperature	< 1%	< 1%
Dissolved oxygen control point	1.89%	1.72%
Conversion factor alpha (α)	5.66%	5.17%
Conversion factor beta (β)	9.43%	12.07%
Standard oxygen transfer efficiency (SOTE)	13.21%	18.86%
Surface aeration standard oxygen transfer rate (R_{std})	20.75%	20.69%

Table 15: Sensitivity analysis results

From the table above it is evident which parameters has the biggest impact on the VTP;'s. It is evident that the "Standard oxygen transfer efficiency (SOTE)" and "Surface aeration standard oxygen transfer rate (R_{std})" parameters have a major effect on the VTP's (thus sensitive parameters) while the "Maximum water temperature" and "Dissolved oxygen control point" is much less sensitive. Special care should therefore be given when choosing or stipulating the values of the most sensitive parameters.

5. CONCLUSION

The main focus of this investigation was to give designers and investors a guide to select the correct aeration technology (SSSA vs FBDA) based on the overall viability of the project. By comparing the cost of these two different aeration technologies with varying input data, it was possible to calculate the Viability Threshold Points (VTPs) for these systems. With the VTPs known, it is possible designers and investors to make calculated decisions on the aeration technology to be used.

From discussions with clients and contractors during this investigation, it would seem there is a reluctance from clients to implement technologies with higher initial capital expenditure, such as the FBDA technology. Especially in smaller municipalities or working with non-technical clients, the capital cost of projects plays such a major role that operational cost is neglected from the overall costing calculations. Convincing clients without actual overall costing figures are difficult and this is where the information represented in this book can hopefully help.

It is however important to remember that the results in this book is based on typical inflow data and designers should give special care when using this information. Designers should therefore carefully study the inflow data used in this investigation and establish where their design inflow data slots in and decide if any adjustments are required. Typical parameters used and assumptions made, like cost functions for specific infrastructure and equipment, design models used, and operating procedures of the plants must be taken into account when using the information. The following disclaimers must be incorporated when using the information:

- Typical key technical design parameters were used for the wastewater characterisation. Should these parameters change considerably, the VTP's will also change. The sensitivity analysis done on the key technical design parameters (section 4.5) should give the designers and investors an indication of the sensitivity of each parameter.
- Although special care has been taken to calculate the costing of associated infrastructure and equipment of the aeration technology as accurately as possible, the costing was done by using cost functions based on historical data and specific infrastructure and equipment. Should the infrastructure and equipment chosen by the designer change considerably, the VTPs will also change. It is therefore recommended that future investigations focus on gathering more cost data of previous projects and thereby increase the accuracy of the cost functions that can be used.

- As explained in the book, the design of the infrastructure and equipment was based on Steady State models. Although the Steady State models gives accurate sizing of the most infrastructure, the sizing of the aeration equipment might be slightly skew depending on the variation in inflow rates and wastewater characteristics encountered on the specific WWTP. It is therefore recommended that for future investigations, a dynamic model simulation be performed on the same data so that a comparison can be made and the results adjusted where required.

- For the calculation of the electrical operating cost (electricity used), it was assumed that the plant will be operated continuously on a 24-hour, 365 day a year basis, running at the average energy consumption calculating. In reality this will not be the case, as a variation in inflow and load will also vary the energy requirements on site. As explained in section 2.7.1, the aeration control system will automatically adjust to keep the dissolved oxygen levels at an optimum level. In this book a reduction of 15% was allowed for on the FBDA technology to account for this adjustment. With more accurate inflow information available and incorporating dynamic models into the calculation, it would be possible to calculate a more accurate electrical operating cost for a specific WWTP. It is therefore encouraged that future investigations focus on gathering more accurate inflow data and electrical consumption data from actual WWTPs in order to get a better relationship between these parameters.

Looking at the disclaimers above it is evident that the results and information must be used with special care and that it will not be applicable in all design situations. The results do however give the designer a benchmark to work from. With the benchmark established, the designer can decide if more accurate cost calculations are required to make the final decision.

What is interesting to note from the results are the low flows at which the VTPs occurs. Although the VTPs might not be 100% accurate in all situations, it does raise the issue that FBDA technology does become viable for much smaller WWTPs than originally estimated. This realisation, backed with the figures in this book, will hopefully give designers the evidence to convince investors and clients to look at more efficient aeration technologies even for smaller plants.

With more and more emphasis being exerted worldwide on reducing humanity's carbon footprint, it is the responsibility of each engineer, investor and client to reduce energy consumption in the design and implementation of projects. The information represented in this book can hopefully assist in this quest and ensure that the correct aeration technology be chosen for each WWTP. This will not only be more financially viable for the client but might also go a long way in reducing our carbon footprint and leave a better future for the next generations to enjoy.

9 783384 226730